Chemistry Problems

Fifth Edition

David E. Newton

J. WESTON
WALCH
PUBLISHER
Portland, Maine

User's Guide
to
Walch Reproducible Books

As part of our general effort to provide educational materials that are as practical and economical as possible, we have designated this publication a "reproducible book." The designation means that purchase of the book includes purchase of the right to limited reproduction of all pages on which this symbol appears:

Here is the basic Walch policy: We grant to individual purchasers of this book the right to make sufficient copies of reproducible pages for use by all students of a single teacher. This permission is limited to a single teacher and does not apply to entire schools or school systems, so institutions purchasing the book should pass the permission on to a single teacher. Copying of the book or its parts for resale is prohibited.

Any questions regarding this policy or requests to purchase further reproduction rights should be addressed to:

Permissions Editor
J. Weston Walch, Publisher
321 Valley Street • P.O. Box 658
Portland, Maine 04104-0658

1   2   3   4   5   6   7   8   9   10

ISBN 0-8251-4266-0

Copyright © 1984, 1993, 2001 David E. Newton
Published by J. Weston Walch
P.O. Box 658 • Portland, Maine 04104-0658
www.walch.com

Printed in the United States of America

Dedicated to

Rick Kimball

*for many years of thoughtful and
constructive help as editor and for
being a good friend*

A Biographical Note

David E. Newton taught math and science at the middle and high school levels in Grand Rapids, Michigan, for 13 years; taught chemistry, physical science, human sexuality, and methods courses at Salem State College in Massachusetts for 15 years; and was Adjunct Professor at the University of San Francisco for 10 years, where he taught courses in science and social issues.

Newton earned his B.S. in chemistry and M.A. in education at the University of Michigan, and his Ed.D. in science education at Harvard University. He has been writing non-fiction for pre-college and general readers for 45 years and has more than 400 titles to his credit. His most recent books include *Social Issues in Science and Technology: An Encyclopedia* (ABC-CLIO), *Chemistry* and *Physics*, both in the Oryx Frontiers of Science Series and the *Encyclopedia of Fire* (Oryx). He is also involved in the development of teacher and student materials for major high school chemistry texts.

Newton is currently co-owner of a nine-room bed-and-breakfast inn in Ashland, Oregon, while continuing to be a "full-time" science writer.

Contents

Appendices

Preface to the Fifth Edition

We are delighted that *Chemistry Problems* continues to be of use to high school teachers throughout the United States and Canada. This publication is now in its fifth decade and still appealing to chemistry teachers everywhere.

This fifth edition contains a number of additions and changes. New appendices have been added that contain chemistry-related web sites and calculator assistance for students working with scientific notation and logarithms. In addition, new problems have been added in the areas of acid and base chemistry and thermochemistry. Finally, the entire text has been brought up to date with the use of currently accepted scientific notation, abbreviations, units, and measurements.

We continue to appreciate and rely on the input from readers who have suggestions as to changes, corrections, and improvements in the book. Please feel free to contact the publisher regarding any suggestions you may have for future revisions.

The author wishes to express his special appreciation to Brian Pressley, responsible for the latest revision of *Chemistry Problems*. Brian has been teaching chemistry and physics since 1993 and is currently employed at Brunswick High School in Maine.

—David E. Newton
Ashland, OR

Exponential Notation

Group I. Express the following numbers using exponential notation.

1. 10 000

2. 0.000 1

3. 10 000 000 000

4. 50 000

5. 2 000 000 000

6. 0.000 000 4

7. 0.000 3

8. 790 000

9. 26 000 000 000 000

10. 0.000 000 045

11. 48 000 000

12. 0.000 000 000 000 000 001 3

13. 0.017 5

14. 0.000 046 2

15. 3 000 000

16. 9 200 000

17. 371

18. 0.009 00

19. 0.62

20. 0.000 197

21. 328 500

22. 76 450

23. 0.941 0

24. 3 005

25. 0.000 705

26. 80 000 000 000

27. 6 240

28. 17 500 000

29. 0.053 02

30. 0.000 000 000 009 15

31. 28

32. 7.480

33. 280 500 000

34. 134.2

35. 0.000 089

Group II. Write the following numbers in the "long form."

1. 1.00×10^5

2. 1×10^{-12}

3. 4×10^2

4. 5×10^{-6}

(continued)

Chemistry Problems

5. 3.2×10^{-2}

6. 7.95×10^{-1}

7. 6.854×10^{14}

8. 14.3×10^2

9. 9.065×10^{-4}

10. 4.3×10^3

11. 44.2×10^7

12. 3.08×10^{-3}

13. 16.9×10^0

14. 66.3×10^4

15. 4.00×10^{-6}

Group III. Carry out the following operations, using exponential notation.

1. $6.000 \times 0.004 =$

2. $50 \times 0.000\ 000\ 2 =$

3. $800 \div 40 =$

4. $4\ 000 + 80 + 500 =$

5. $30\ 000 \times 450\ 000 =$

6. $0.006 \div 0.2 =$

7. $0.000\ 2 \times 0.04 =$

8. $32\ 000 \div 400\ 000 =$

9. $1\ 000\ 000 \times 7\ 900 =$

10. $60\ 000 + 7\ 000 + 80\ 000 =$

11. $0.000\ 045 \div 0.009 =$

12. $0.002\ 4 \times 20\ 000 =$

13. $0.006\ 4 \div 8\ 000 =$

14. $0.090 \div 0.000\ 03 =$

15. $0.000\ 4 \times 5\ 000 \times 0.02 =$

16. $450 + 250 + 7\ 500 =$

17. $400 \div 200\ 000 =$

18. $6 \div 30\ 000 =$

19. $90 \times 0.02 \times 0.000\ 01 =$

20. $0.000\ 081 \div 90\ 000 =$

21. $40\ 000\ 000 \times 0.001 \times 0.20 =$

22. $300\ 000\ 000 \div 0.000\ 6 =$

23. $6\ 450 + 28\ 200 + 3\ 740 =$

24. $0.060 \times 20\ 000 \times 0.000\ 2 =$

25. $4 \div 0.000\ 000\ 2 =$

26. $13 \times 0.000\ 001 \times 4\ 000 =$

27. $4\ 850 - 370 =$

28. $0.006 \times 0.000\ 2 \times 0.02 =$

29. $0.008\ 4 \div 42 =$

30. $1\ 400 \times 2\ 000\ 000 \times 0.004 =$

31. $0.004\ 5 + 0.055\ 0 + 0.001\ 5 =$

32. $0.000\ 72 \div 900\ 000 =$

33. $0.000\ 5 \times 200 \times 40\ 000\ 000 =$

34. $0.049 - 0.000\ 17 =$

(continued)

Name _____

Date _____

35. $0.000\ 2 \times 8\ 000 \times 0.02 \times 2\ 000 =$

36. $\dfrac{80\ 000\ 000 \times 140}{0.04} =$

37. $\dfrac{640 \times 0.072 \times 4}{160\ 000 \times 0.000\ 36} =$

38. $\dfrac{30}{60\ 000 \times 0.05} =$

39. $\dfrac{42 \times 0.007\ 0}{0.35 \times 30\ 000 \times 140} =$

40. $\dfrac{450 \times 0.008 \times 0.02}{20\ 000 \times 40 \times 0.18} =$

41. $\dfrac{28\ 000 \times 0.005\ 6}{0.000\ 49 \times 16} =$

42. $\dfrac{240\ 000 \times 0.003}{0.000\ 16} =$

43. $\dfrac{0.000\ 035}{700 \times 0.001} =$

44. $\dfrac{32 \times 0.000\ 7 \times 0.02}{20\ 000 \times 0.04 \times 70} =$

45. $\dfrac{150\ 000 \times 0.008 \times 200}{0.05 \times 160} =$

46. $\dfrac{400\ 000 \times 0.000\ 003\ 6}{30 \times 0.000\ 000\ 2 \times 1.2} =$

47. $\dfrac{420 \times 0.990\ 0 \times 80 \times 0.000\ 5}{200 \times 30\ 000 \times 3.6} =$

48. $\dfrac{350\ 000\ 000 \times 0.006\ 4 \times 2\ 400}{1.6 \times 0.000\ 14 \times 36\ 000 \times 0.000\ 005} =$

49. $\dfrac{4.9 \times 10\ 000 \times 450\ 000 \times 8.0}{0.005 \times 6\ 000 \times 0.000\ 72 \times 0.3} =$

50. $\dfrac{2.8 \times 0.000\ 002\ 5 \times 0.000\ 000\ 14}{0.000\ 7 \times 50\ 000 \times 0.002 \times 210\ 000} =$

Significant Figures and Errors of Measurement

Name _____

Date _____

Group I. Determine the number of significant figures in each of the following numbers.

1. 5.432 g

2. 40.319 g

3. 146 cm^3

4. 3.285 cm

5. 0.189 kg

6. 429.3 g

7. 2873.0 cm^3

8. 99.9 cm^3

9. 0.000 235 g

10. 144 kg

11. 2500 cm

12. 2500.0 cm

13. 1.04×10^{14} g

14. 3.58×10^{-9} nm

15. 48.571 93 kg

16. 8365.6 g

17. 0.002 300 mg

18. 7.500×10^8 cg

19. 3.92×10^{-4} g

20. 1.000×10^3 kg

Group II. Add the following.

1. 12 cm + 0.031 cm + 7.969 cm =

2. 0.085 cm + 0.062 cm + 0.14 cm =

3. 3.419 g + 3.912 g + 7.051 8 g + 0.000 13 g =

4. 30.5 g + 16.82 g + 41.07 g + 85.219 g =

5. 143.0 cm + 289.25 cm + 68.45 cm + 6.00 cm =

6. 29.49 cm + 83.46 cm + 107.05 cm + 26.618 cm =

7. 0.065 3 g + 0.085 38 g + 0.076 54 g + 0.043 2 g =

8. 1.8×10^{-5} cm + 3.25×10^{-4} cm + 4.6×10^{-5} cm =

9. 63.489 cm^3 + 126.2 cm^3 + 68.85 cm^3 + 12.05 cm^3 =

10. 2.3×10^2 g + 4.62×10^2 g + 3.852×10^2 g =

(continued)

Name _____

Date _____

Group III. Subtract the following.

1. 41.025 cm – 23.38 cm =

2. 289 g – 43.7 g =

3. 145.63 cm^3 – 28.9 cm^3 =

4. 62.47 g – 39.9 g =

5. 40.008 cm^3 – 29.094 1 cm^3 =

Group IV. Multiply the following.

1. 2.89 cm × 4.01 cm =

2. 17.3 cm × 6.2 cm =

3. 3.08 m × 1.2 m =

4. 5.00 mm × 7.321 6 mm =

5. 20.8 dm × 123.1 dm =

6. 5 cm × 5 cm =

7. 5.0 cm × 5 cm =

8. 5.0 cm × 5.0 cm =

9. 4.8×10^2 m × 2.101×10^3 m =

10. 9.13×10^{-4} cm × 1.2×10^{-3} cm =

11. 4.218 cm × 6.5 cm =

12. 150.0 m × 4.00 m =

13. 282.2 km × 3.0 km =

14. 14×10^{-8} m × 3.25×10^{-6} m =

15. 2.865×10^4 m × 1.47×10^3 m =

Group V. Divide the following.

1. 8.071 cm^2 ÷ 4.216 cm =

2. 109.375 8 m^2 ÷ 5.813 m =

3. 24,789.4 km^2 ÷ 43.5 km =

4. 6.058 mm^2 ÷ 0.85 mm =

5. 4.819 cm^2 ÷ 9.852 cm =

6. 139.482 m^2 ÷ 68.75 m =

7. 4.23 m^2 ÷ 18.941 m =

8. 85.621 km^2 ÷ 8.05 km =

9. 6.023×10^{14} mm^2 ÷ 5.813×10^{12} mm =

10. 1.142×10^{-8} mm^2 ÷ 8.5×10^{-4} mm =

(continued)

Chemistry Problems

Group VI. Express the answers to the following problems using significant figures.

1. $0.057 \text{ cm}^3 \times \dfrac{760 \text{ mmHg}}{740 \text{ mmHg}} \times \dfrac{273 \text{ K}}{250 \text{ K}} =$

2. $142.0 \text{ cm}^3 \times \dfrac{745 \text{ mmHg}}{785 \text{ mmHg}} \times \dfrac{300.0 \text{ K}}{295 \text{ K}} =$

3. $51.3 \text{ g} \times \dfrac{44.962 \text{ amu}}{115.874 \text{ amu}} =$

4. $83.495 \text{ g} \times \dfrac{172.76 \text{ g}}{260.00 \text{ g}} =$

5. $6.025 \times 10^{14} \text{ cm}^3 \times \dfrac{20\ 000.0 \text{ mmHg}}{142.5 \text{ mmHg}} \times \dfrac{273.0 \text{ K}}{315.0 \text{ K}} =$

Group VII. Solve each of the following problems involving errors of measurement.

1. As the result of experimental work, a student finds the density of a liquid to be 0.1369 g/cm^3. The known density of that liquid is 0.1478 g/cm^3. What are the absolute and percent errors of this student's work?

2. After heating a 10.00 g sample of potassium chlorate, a student obtains an amount of oxygen calculated to be 3.90 g. Theoretically, there should be 3.92 g of oxygen in this amount of potassium chlorate. What are the absolute and percent errors in this experiment?

3. The melting point of potassium thiocyanate determined by a student in the laboratory turns out to be 174.5 °C. The accepted value of this melting point is 173.2 °C. What are the absolute and percent errors in this experimental result?

4. A person attempting to lose weight on a diet weighs 80 kg on a bathroom scale at home. An hour later at the doctor's office, on a more accurate scale, this person's weight is recorded as 81 kg. Assuming that there was no real weight change in that hour, what are the absolute and percent errors between these readings?

(continued)

5. The theoretical yield predicted in a particular chemical reaction is 0.106 2 g. The actual yield obtained by a chemist in a specific experiment is 0.009 8 g. Calculate the absolute and percent errors for this experiment.

6. A student decides it is possible to estimate the capacity of a test tube by treating it as a rectangle and neglecting its "roundness." On this basis, the student finds the capacity to be 100.5 cm^3. In fact, the real capacity of the test tube is 100.0 cm^3. What percent error has resulted from the student's neglecting the "roundness" of the test tube?

7. A chemist attempts to determine the surface tension of various detergent-containing liquids by using a tensiometer. In determining the accuracy of the instrument, the chemist tests the surface tension of pure water and obtains a value of 71.28 dynes/cm. The standard value for this quantity is 71.97 dynes/cm. What are the absolute and percent errors of this tensiometer?

8. In an exercise to teach students how to use an analytical balance, the instructor gives a student a quarter which was pre-massed as 5.602 6 g. The mass that the student obtains for the same quarter is 5.601 3 g. What are the absolute and percent errors in this student's weighing?

9. The concentration determined for an unknown sample of hydrochloric acid by a student is 0.135 5 N. According to the instructor's information, the true normality of this solution is 0.136 4 N. What are the absolute and percent errors in this titration?

10. An object with a mass of exactly (and correctly) 0.54 g is given to two students. One student obtains a mass of 0.59 g for the object, while another says the mass is 0.49 g. Which of the students, if either, has the greater percent error in this experiment?

11. In an experiment to determine the density of a liquid, an error of no more than 5.0% is permitted. The true value for this density is 1.439 2 g/cm^3. What are the maximum and minimum values which a student may obtain to fall within the acceptable range?

12. An instructor has decided that students' weighing of a metal bar must lie within ± 2.5% of the bar's true mass of 84.556 g. What is the range of masses which will be accepted?

(continued)

13. What are the minimum and maximum possible values of the measurement 865.8 ± 0.4 cm?

14. What is the uncertainty of the result of adding 35.08 ± 0.02 g and 40.05 ± 0.01 g?

15. What is the uncertainty in the product of 350 ± 4 cm and 55 ± 3 cm?

16. What is the smallest number that could result from subtracting 22 ± 2 from 35 ± 3?

17. The dimensions of a rectangle are measured to be 20.0 ± 0.1 cm and 2.5 ± 0.1 cm. Express the area of the rectangle, including the uncertainty of the measurement.

18. 237.5 ± 0.01 g of a liquid have a volume of 1 500 ± 5.0 cm³. Calculate the density of this liquid, showing the uncertainty in the answer.

19. A chemist precipitates out a certain amount of a chemical into a crucible whose mass is 55.127 ± 0.001 g. After precipitation, the crucible and contents have a mass of 73.641 ± 0.001 g.

    (a) Calculate the amount of precipitate formed.

    (b) Calculate the uncertainty in this measurement.

20. In a particular experiment, the heat from a burning candle is used to raise the temperature of a volume of water. The results of the experiment show that 1.24 ± 0.02 g of the candle are consumed, while 430 ± 2.0 g of water has its temperature raised by 54.0 ± 0.2 °C. What amount of heat was absorbed by the water? Indicate the uncertainty of your answer.

Dimensional Analysis

Name _____

Date _____

Use dimensional analysis to convert each of the following measurements to its equivalent in the units given.

1. 14 cm to meters

2. 31 g to milligrams

3. 116.5 m to kilometers

4. 285.9 cm to kilometers

5. 0.006 394 km to centimeters

6. 8.4×10^{-6} kg to centigrams

7. 1.47×10^5 mm to kilometers

8. 4.7 kg to centigrams

9. 138.4 mg to grams

10. 65.5 km to meters

11. 23.6 dm to centimeters

12. 2.36×10^4 s to days

13. 13.6 dm^3 to cubic centimeters

14. 20.6 km/hr to meters per second

15. 0.058 m/s to centimeters per second

16. 3.49 km/hr to meters per second

17. 14.7 g/cm^3 to centigrams per cubic millimeter

18. 7.3×10^{-4} cm^3/s to cubic centimeters per day

19. 8.05×10^5 g/cm^3 to kilograms per dm^3

20. 3.42×10^3 kg/m^2 to grams per square centimeter

Metric System

Name _____

Date _____

Group I. Convert the following measurements to milligrams.

1. 73 g

2. 165.3 dg

3. 0.002 38 g

4. 1.06 cg

5. 0.042 dag

6. 3.01×10^{-6} cg

7. 41.05 cg

8. 0.2 kg

9. 13.52 g

10. 4.6×10^{-4} g

11. 3.4 dg

12. 0.52 µg

Group II. Convert the following measurements to centigrams.

1. 938 mg

2. 19.3 g

3. 228 mg

4. 14.3 kg

5. 243.2 g

6. 0.28 dg

7. 8.5 dg

8. 0.31 kg

9. 6.3×10^4 mg

10. 8.1 g

11. 8.012×10^{14} µg

12. 4.2 hg

Group III. Convert the following measurements to decigrams.

1. 73.456 cg

2. 36.9 mg

3. 0.041 g

4. 1.52 g

5. 124.0 mg

6. 142.81 g

7. 0.895 kg

8. 23 cg

9. 73 g

10. 238 µg

(continued)

Chemistry Problems

Group IV. Convert the following measurements to grams.

1. 41.5 cg

2. 3.89 mg

3. 0.978 5 kg

4. 9.9 kg

5. 11.9 dag

6. 28.5 cg

7. 75.3 cg

8. 14.587 mg

9. 28.1 dg

10. 2.3 dag

11. 423 mg

12. 13.4 mg

Group V. Convert the following measurements to kilograms.

1. 2450 g

2. 147.2 hg

3. 89.9 g

4. 23.8 μg

5. 0.050 cg

6. 283 mg

7. 65.4 dag

8. 2.4×10^6 dg

9. 1.47×10^3 g

10. 6×10^{-4} mg

Group VI. Convert the following measurements to kilograms.

1. 14 g

2. 135.4 g

3. 71.9 mg

4. 5.1×10^4 g

5. 58.3 hg

6. 14.2 mg

7. 62 cg

8. 0.52 dg

9. 3.9 dag

10. 4.2 mg

Group VII. Convert the following measurements to grams.

1. 23.7 dag

2. 0.087 dg

3. 329 mg

4. 1.07 kg

5. 39.82 mg

6. 1.2 μg

(continued)

Name _____

Date _____

7. 4.3 kg

8. 46.5 cg

9. 3.0×10^4 mg

10. 74.3 cg

Group VIII. Convert the following measurements to cubic decimeters.

1. 182 mL

2. 3.4 hm^3

3. 4.2 km^3

4. 0.895 cm^3

5. 135.4 mL

6. 750 hm^3

7. 22.9 cm^3

8. 43.2 mm^3

9. 158 hm^3

10. 28 dam^3

11. 60 dam^3

12. 250 cm^3

Group IX. Convert the following measurements to milliliters.

1. 14.2 L

2. 2.590 1 mm^3

3. 9.23×10^{-6} dam^3

4. 7.20 dm^3

5. 0.323 L

6. 148 dam^3

7. 0.416 mm^3

8. 4.38 hm^3

9. 243 dm^3

10. 1.079×10^{-12} dam^3

Group X. Convert the following measurements to cubic decimeters.

1. 14.3 mm^3

2. 254 cm^3

3. 70.5 m^3

4. 55.5 cm^3

5. 3.0 mm^3

6. 386.9 m^3

7. 6.03×10^4 hm^3

8. 128 mm^3

9. 9.00 cm^3

10. 11.2 dam^3

(continued)

Group XI. Convert the following measurements to millimeters.

1. 16 μm
2. 85.4 cm
3. 4.32×10^{-8} m
4. 5.8×10^{-2} μm
5. 72.0 dm
6. 125 μm
7. 234.5 km
8. 14.2 dm
9. 31.9 cm
10. 7.68 m
11. 6.0×10^{4} Å
12. 3.04 dm

Group XII. Convert the following measurements to centimeters.

1. 3.67 dm
2. 124 mm
3. 9.4 μm
4. 6.00 dam
5. 2.0×10^{4} mm
6. 36 dm
7. 11 μm
8. 85.9 m
9. 0.85 m
10. 113 hm
11. 7.4×10^{-5} m
12. 1 047 Å

Group XIII. Convert the following measurements to decimeters.

1. 14.5 m
2. 5.15 mm
3. 57 m
4. 14.4 cm
5. 3.21×10^{11} μm
6. 1.4×10^{-4} cm
7. 106 μm
8. 6.7×10^{-8} hm
9. 0.0628 mm
10. 42.0 m

Group XIV. Convert the following measurements to meters.

1. 1.0 km
2. 1.0×10^{-9} mm
3. 78.5 cm
4. 39.0 dm

(continued)

Chemistry Problems

5. 1.00 cm

6. 34.4 mm

7. 0.468 km

8. 0.70 dm

9. 6 808 Å

10. 0.99 hm

Group XV. Convert the following measurements to kilometers.

1. 23.896 cm

2. 562.0 μm

3. 4.12×10^6 mm

4. 181 m

5. 6.7 hm

6. 37.1×10^8 dm

7. 7.4 dam

8. 7.63 m

9. 0.023 cm

10. 43.5 m

Group XVI. Convert the following measurements to centimeters.

1. 245 mm

2. 0.438 m

3. 41.5 μm

4. 71.0 mm

5. 6.4 dm

6. 18.744 dm

7. 8.3 m

8. 379 μm

9. 3.11 m

10. 2438 Å

Group XVII. Convert the following measurements to decimeters.

1. 5.67 m

2. 2.80 cm

3. 70.58 hm

4. 4.03×10^{-4} km

5. 5.07 m

6. 4.0 mm

7. 31.43 hm

8. 6.21×10^8 μm

9. 67.3 mm

10. 121 cm

(continued)

Group XVIII. Convert the following measurements to kilometers.

1. 13 m
2. 65.0 dm
3. 27.0 hm
4. 3.06 cm
5. 28.9 m

6. 2 180 dm
7. 413.3 nm
8. 745 hm
9. 0.051 cm
10. 16.04 dam

Group XIX. Convert the following measurements to micrometers (microns).

1. 1.00 mm
2. 329. cm
3. 0.058 dm
4. 3.5 cm
5. 11 075 Å

6. 4.0 dm
7. 4 812.0 Å
8. 6.9×10^{-12} mm
9. 4.28×10^{-6} cm
10. 129 mm

Group XX. Change the following measurements to angstroms.

1. 63.4 mm
2. 6.4×10^{-6} cm
3. 4.017 cm
4. 1 080.0 µm
5. 75.0 mm

6. 3.0×10^{8} µm
7. 1.00 km
8. 9.0×10^{-3} mm
9. 38.2 dm
10. 111.1 µm

Group XXI. Convert the odd-numbered problems below to picograms, the even-numbered problems to nanograms.

1. 4.5 g
2. 0.008 2 g
3. 0.065 mg

4. 3.17 mg
5. 2.9×10^{-4} g
6. 1.5×10^{-6} g

(continued)

Chemistry Problems

7. 1.23×10^{-4} mg

8. 8.85×10^{-3} µg

9. 3.6×10^{-3} ng

10. 4.28×10^5 pg

11. 28.6 ng

12. 153 pg

Group XXII. Solve the following problems.

1. A box having the dimensions 14 cm by 8.0 cm by 3.0 cm is filled with water. What mass does this amount of water have?

2. A cylindrical tube has a volume of 30.0 cm^3. Its diameter is 3.0 cm. What is its height?

3. What are the dimensions of a cube holding 250.0 g of water?

4. A test tube is 13.0 cm long and 3.0 cm in diameter. What is its volume in cubic centimeters?

5. Sixteen students each take 25 cm^3 of nitric acid from a stock bottle holding 1 dm^3 of nitric acid. How much acid remains?

6. Determine the equivalent distance in kilometers of the running distance of 400.0 m.

7. A service station attendant sells 1.5 dm^3 of gasoline to each of four chemistry students. If this gasoline was taken from a 50-dm^3 container, how much gasoline remains in the tank?

8. A long thin glass tube has an average internal diameter of 1.5×10^{-4} cm and a total length of 75 cm. What is the volume of this tube?

(continued)

9. A large river flows at the rate of 2.6×10^5 dm^3/s into a boat lock 75 m long, 30.0 m wide, and 18 m high. How long before the lock will be filled with water?

10. A micropipette holds a total volume of 0.10 cm^3. The length of the instrument is 25 cm. What is its internal diameter?

11. A large volumetric flask has essentially a spherical shape with a diameter of 12.5 cm. What is its exact volume?

12. The length and width of a container are, respectively, 18 cm and 15 cm. If the container holds 6.0 kg of water, what is the height of the container?

13. Twelve samples of cupric sulfate, weighing 30.0 g each, are taken from a one-kilogram bottle. How many grams of cupric sulfate are left?

14. A burette delivers 0.15 cm^3 of water per second. How much time is required for the burette to transfer 18 g of water?

15. A one-liter bottle of carbon tetrachloride is used to fill a number of vials. If 305 cm^3 of carbon tetrachloride remain when all the vials are filled, how much carbon tetrachloride has been used?

Energy

Name _____

Date _____

Group I. Convert each of the following temperature readings to its equivalent in degrees Celsius.

1. 32 K

2. 58 K

3. 5.8 K

4. 139.6 K

5. 42.4 K

6. 7.9 K

7. 273 K

8. 313 K

9. 330.4 K

10. 406.3 K

11. 134.7 K

12. 48.9 K

Group II. Convert each of the following temperature readings to its equivalent in Kelvin.

1. 100 °C

2. 0 °C

3. 26 °C

4. 175.4 °C

5. –40.0 °C

6. –155.3 °C

7. 308 °C

8. 361 °C

9. 285.1 °C

10. 196.4 °C

11. 416.5 °C

12. 89.3 °C

Group III. Convert each of the following temperature readings to its equivalent in Kelvin.

1. 10 °C

2. 78 °C

3. 35.4 °C

4. 266.3 °C

5. 159.5 °C

6. –64.3 °C

7. 17 °C

8. 106.4 °C

9. 63.7 °C

10. 214.5 °C

11. –38.7 °C

12. –262.2 °C

(continued)

Group IV. Determine the amount of heat (in joules) absorbed or released in each of the following changes.

1. 40.0 g of water heated from 10.0 °C to 30.0 °C

2. 25.0 g of water cooled from 85.0 °C to 40.0 °C

3. 65.5 g of water heated from 32.5 °C to 48.7 °C

4. 135.6 g of water cooled from 95.8 °C to 21.6 °C

5. 100.0 g of ice melted, with no temperature change

6. 40.0 g of water, boiled at 100.0 °C

7. 30.0 g of aluminum heated from 15.0 °C to 35.0 °C

8. 450.0 g of iron cooled from 125.0 °C to 45.0 °C

9. 62.3 g of lead heated from 21.7 °C to 136.4 °C

10. 195.4 g of magnesium cooled from 120.6 °C to 14.9 °C

11. 1.5 kg of copper heated from 5.5 °C to 132.0 °C

12. 47.8 g of lead melted, with no change in temperature

13. 186.3 g of copper changed from liquid to solid state at its melting point

14. The complete vaporization of 53.8 g of lead at its boiling point

15. The condensation from vapor to liquid state of 235.5 g of copper at its boiling point

16. The heating of a mixture of 5.8 g of lead and 6.2 g of copper from 5.8 °C to 12.1 °C

(continued)

17. The heating of 125.6 g of ice from –165.8 °C to –38.6 °C

18. The cooling of 145.5 g of steam from 382.5 °C to 107.7 °C

19. The complete vaporization of 10.0 g of ice originally at –10.0 °C

20. The complete vaporization of 35.8 g of ice originally at –65.4 °C

Group V. Solve the following problems involving energy changes.

1. To change the temperature of a particular calorimeter and the water it contains by one degree Celsius requires 6485 J. The combustion of 2.80 g of ethylene gas, C_2H_4, in the calorimeter causes a temperature rise of 21.4 °C. From this information, find the heat of combustion, per mole, of ethylene.

2. Which of the following reactions are endothermic?

    (a) H_2 (g) + S (s) + $2O_2$ (g) $\rightleftharpoons$ H_2SO_4 (l) $\Delta H° = -811.3$ kJ

    (b) 3C (s) + $2Fe_2O_3$ (s) + 463.6 kJ $\rightleftharpoons$ 4Fe (s) + $3CO_2$ (g)

    (c) H_2O (l) $\rightleftharpoons$ H_2 (g) + $\frac{1}{2}O_2$ (g) $\Delta H° = +285.9$ kJ

    (d) C (s) + $2H_2$ (g) $\rightleftharpoons$ CH_4 (g) + 74.9 kJ

3. The heat of fusion for ice is 334 kJ/kg. How many joules would be needed to melt 7 mol of ice?

4. For a certain chemical reaction, at 400 K, $\Delta G = -73.6$ kJ, and $\Delta H = -110.9$ kJ. Find the entropy change for this process at this temperature.

5. Given:

    (a) Na (s) + $\frac{1}{2}Cl_2$ (g) $\rightarrow$ NaCl (s) $\Delta H° = -411$ kJ

    (b) H_2 (g) + S (s) + $2O_2$ (g) $\rightarrow$ H_2SO_4 (l) $\Delta H° = -811.3$ kJ

    (c) 2Na (s) + S (s) + $2O_2$ (g) $\rightarrow$ Na_2SO_4 (s) $\Delta H° = -1383$ kJ

    (d) $\frac{1}{2}H_2$ (g) + $\frac{1}{2}Cl_2$ (g) $\rightarrow$ HCl (g) $\Delta H° = -92.3$ kJ

(*continued*)

20

Find the heat of reaction of the following chemical change:

$$2NaCl \ (s) + H_2SO_4 \ (l) \rightarrow Na_2SO_4 \ (s) + 2HCl \ (g)$$

6. Calculate ΔH_{25}° for the reaction, and tell whether the reaction is exothermic or endothermic. Given the following values for ΔH_f°:

   $Na \ (s) = 0$ $H_2 \ (g) = 0$

   $NaOH \ (s) = -426.8 \ kJ/mol$ $H_2O \ (l) = -285.8 \ kJ/mol$

   $2Na \ (s) + 2H_2O \ (l) \rightarrow 2 \ NaOH \ (s) + H_2 \ (g)$

7. How much heat energy (in joules) is given off when one kilomole of hydrogen gas (H_2) at 25 °C and 1 atmosphere is combined with enough oxygen (O_2) to make liquid water at 25 °C?

8. Predict the heat of reaction for the following reaction:

   $$CO \ (g) + \tfrac{1}{2}O_2 \ (g) \rightarrow CO_2 \ (g)$$

   (Consult tables for any data required in this problem.)

9. Calculate the heat of combustion for the reaction in which ethane combines with oxygen to give carbon dioxide and water vapor. Data to which you may refer include the following:

   $C_2H_6 \ (g) \rightarrow 2C \ (s) + 3H_2 \ (g)$ $\Delta H = 84.5 \ kJ$

   $C \ (s) + O_2 \ (g) \rightarrow CO_2 \ (g)$ $\Delta H = -393 \ kJ$

   $H_2 + \tfrac{1}{2}O_2 \ (g) \rightarrow H_2O \ (g)$ $\Delta H = -242 \ kJ$

10. Calculate the entropy change for (a) the melting of ice at 0 °C and (b) the vaporization of water at 100 °C.

11. What is the change in free energy at 25 °C for the reaction in which hydrogen and carbon dioxide react to form water vapor and carbon monoxide?

12. Water gas (hydrogen and carbon monoxide) is formed when steam is passed over hot charcoal. Calculate the free energy for this reaction at 25 °C.

(continued)

Name _____

Date _____

Group VI. Solve the following enthalpy-change problems.

1. How much heat will be transferred when 5.81 g of graphite reacts with excess H_2 as follows? Is this reaction endothermic or exothermic?

   $$6C + 3H_2 \rightarrow C_6H_6 \qquad \Delta H° = 49.03 \text{ kJ}$$

2. How much heat will be released when 6.44 g of sulfur reacts with excess O_2 as follows?

   $$2S + 3O_2 \rightarrow 2SO3 \qquad \Delta H° = -791.4 \text{ kJ}$$

3. How much heat will be transferred when 14.9 g of ammonia reacts with excess O_2 as follows? Is this reaction endothermic or exothermic?

   $$4NH_3 + 5O_2 \rightarrow 4NO + 6H_2O \qquad \Delta H° = -1170 \text{ kJ}$$

4. How much heat will be released when 4.72 g of carbon reacts with excess O_2 as follows?

   $$C + O_2 \rightarrow CO_2 \qquad \Delta H° = -393.5 \text{ kJ}$$

5. How much heat will be absorbed when 13.7 g of nitrogen reacts with excess O_2 as follows?

   $$N_2 + O_2 \rightarrow 2NO \qquad \Delta H° = 180 \text{ kJ}$$

6. How much heat will be absorbed when 38.2 g of bromine reacts with excess H_2 according to the following equation?

   $$H_2 + Br_2 \rightarrow 2HBr \qquad \Delta H° = 72.80 \text{ kJ}$$

7. How much heat will be released when 11.8 g of iron reacts with excess O_2 according to the following equation?

   $$3Fe + 2O_2 \rightarrow Fe_3O_4 \qquad \Delta H° = -1120.48 \text{ kJ}$$

8. How much heat will be released when 1.48 g of chlorine reacts with excess phosphorus as follows?

   $$2P + 5Cl_2 \rightarrow 2PCl_5 \qquad \Delta H° = -886 \text{ kJ}$$

9. How much heat will be released when 4.77 g of ethanol (C_2H_5OH) reacts with excess O_2 according to the following equation?

   $$C_2H_5OH + 3O_2 \rightarrow 2CO_2 + H_2O \qquad \Delta H° = -1366.7 \text{ kJ}$$

10. How much heat will be released when 18.6 g of hydrogen reacts with excess O_2 as follows?

    $$2H_2 + O_2 \rightarrow H_2O \qquad \Delta H° = -571.6 \text{ kJ}$$

 Chemistry Problems

Atomic Structure

Group I. Draw a labeled diagram of each of the following atoms. Show the number of protons, the number of neutrons, and the probable arrangement of electrons, as far as can be determined.

Z = atomic number   A = mass number

1. oxygen   Z   8   7. potassium   Z   19
                 A   16   A   39

2. lithium   Z   3   8. helium   Z   2
                 A   7   A   4

3. calcium   Z   20   9. aluminum   Z   13
                 A   40   A   27

4. sulfur   Z   16   10. hydrogen   Z   1
                 A   32   A   1

5. argon   Z   18   11. beryllium   Z   4
                 A   40   A   9

6. carbon   Z   6   12. fluorine   Z   9
                 A   12   A   19

(continued)

Name _____

Date _____

13. nitrogen	Z	7		17. ruthenium	Z	44
	A	14			A	101
14. arsenic	Z	33		18. platinum	Z	78
	A	75			A	195
15. bismuth	Z	83		19. cobalt	Z	27
	A	209			A	59
16. gold	Z	79		20. scandium	Z	21
	A	197			A	45

Group II. Write the electron configuration found in each of the following atoms.

1. oxygen

2. chlorine

3. titanium

4. beryllium

5. cobalt

6. calcium

7. strontium

8. nickel

9. silicon

10. rubidium

11. aluminum

12. chromium

13. scandium

14. phosphorus

15. zinc

16. bromine

17. potassium

18. vanadium

19. gallium

20. selenium

(continued)

Atomic Structure *(continued)*

Group III. Write the quantum numbers of all the electrons in the following atoms.

1. boron

2. fluorine

3. sodium

4. argon

5. oxygen

6. helium

7. potassium

8. hydrogen

9. magnesium

10. beryllium

11. lithium

12. carbon

Group IV. Draw diagrams to show how the following isotopes differ from each other.

1. hydrogen-1 and hydrogen-2

2. lithium-6 and lithium-7

3. boron-10 and boron-11

4. chlorine-35 and chlorine-37

5. argon-35, argon-38, and argon-40

6. germanium-70, germanium-72, germanium-73, germanium-74, and germanium-76

(continued)

7. uranium-234 and uranium-238

8. tellurium-122, tellurium-124, tellurium-126, and tellurium-128

9. xenon-129, xenon-130, xenon-131, xenon-132, xenon-134, and xenon-136

10. europium-151 and europium-153

Group V. Calculate the atomic mass of the elements listed in Group IV. Assume the isotopic composition of each is as given below.

1. 99.985% ^{1}H; 0.015% ^{2}H

2. 7.42% ^{6}Li; 92.58% ^{7}Li

3. 19.78% ^{10}B; 80.22% ^{11}B

4. 75.53% ^{35}Cl; 24.47% ^{37}Cl

5. 0.337% ^{35}Ar; 0.063% ^{38}Ar;   99.60% ^{40}Ar

6. 20.52% ^{70}Ge; 27.43% ^{72}Ge; 7.76% ^{73}Ge; 36.54% ^{74}Ge; 7.76% ^{76}Ge

7. 0.70% ^{234}U; 99.30% ^{238}U

8. 0.089% ^{122}Te; 2.46% ^{124}Te; 5.48% ^{126}Te; 91.97% ^{128}Te

9. 28.44% ^{129}Xe; 4.08% ^{130}Xe; 21.18% ^{131}Xe; 26.89% ^{132}Xe; 10.44% ^{134}Xe; 9.01% ^{136}Xe

10. 47.82% ^{151}Eu; 52.18% ^{152}Eu

Group VI. Answer each of the following questions on atomic structure.

1. How many valence electrons does each of the following have?

(a) krypton

(b) arsenic

(c) lithium

(d) fluorine

(e) strontium

(f) silicon

(g) boron

(h) sulfur

(continued)

Chemistry Problems

2. Using the inert gas core convention, write the electron configuration for the following.

   (a) bismuth

   (b) bismuth ion (Bi^{3+})

   (c) xenon

   (d) xenon ion (Xe^-)

   (e) rubidium

   (f) rubidium ion (Rb^+)

   (g) manganese

   (h) manganese ion (Mn^{7+})

3. What element corresponds to each of the following electron configurations?

   (a) $1s^2 2s^2 2p^3$

   (b) $1s^2 2s^2$

   (c) $1s^2 2s^2 2p^6 3s^2 3p^3$

   (d) $1s^2 2s^2 2p^6 3s^2 3p^6 4s^2 3d^5$

   (e) $1s^2 2s^2 2p^6 3s^2 3p^6 4s^1$

   (f) $1s^2 2s^2 2p^6 3s^2 3p^6 4s^2 3d^{10} 4p^6 5s^2 4d^2$

   (g) $1s^2 2s^2 2p^6 3s^2 3p^6 4s^2 3d^{10} 4p^6 5s^2 4d^{10} 5p^6 6s^2 4f^5$

   (h) $1s^2 2s^2 2p^6 3s^2 3p^6 4s^2 3d^{10} 4p^4$

4. Write the electron configuration for each of the following elements in the oxidation states given.

   (a) Cu^0, Cu^{1+}, Cu^{2+}

   (b) Al^0, Al^{3+}

   (c) Cl^0, Cl^{1-}

   (d) O^0, O^{2-}, O^{1-}

   (e) $S^0, S^{2+}, S^{4+}, S^{6+}, S^{2-}$

   (f) $P^0, P^{5+}, P^{3+}, P^{3-}$

5. Using the $\boxed{\uparrow\downarrow}$ convention, draw the electron configurations for each of the following.

   (a) potassium

   (b) helium

   (c) beryllium ion (Be^{2+})

   (d) nitrogen

   (e) fluoride ion (F^{1-})

(continued)

(f) sulfur

(g) silicon ion (Si^{4+})

(h) neon

6. Write the Lewis electron-dot symbol for each of the following.

(a) carbon

(b) aluminum

(c) boron

(d) hydrogen

(e) fluorine

(f) neon

(g) sodium

(h) potassium ion (K^+)

(i) sulfur ion (S^{2-})

(j) magnesium ion (Mg^{2+})

(k) boron ion (B^{3+})

(l) argon ion (Ar^+)

7. Using the $\boxed{\uparrow\downarrow}$ convention, show the electron configuration in the lowest energy state of the following.

(a) oxygen

(b) boron

(c) aluminum

(d) calcium

(e) sodium ion (Na^+)

(f) bromide ion (Br^-)

(g) magnesium ion (Mg^{2+})

(h) sulfide ion (S^{2-})

Bonding Reactions

Name _____

Date _____

Determine the type of bonding to be expected in each of the following cases. Then, use electron-dot symbols to show how the bonding takes place.

1. sodium and chlorine

2. potassium and fluorine

3. lithium and sulfur

4. carbon and hydrogen

5. beryllium and sulfur

6. magnesium and nitrogen

7. silicon and fluorine

(continued)

Name _____

Date _____

8. aluminum and nitrogen

9. carbon and chlorine

10. carbon and fluorine

11. sulfur and oxygen

12. germanium and hydrogen

13. calcium and chlorine

14. magnesium and phosphorus

15. zinc and bromine

Formulas and Nomenclature

Name _____

Date _____

Group I. Name the following compounds.

1. HCl

2. KOH

3. HgOH

4. KCl

5. $FeCl_3$

6. HNO_3

7. NH_4OH

8. Cu_2O

9. $Al_2(SO_4)_3$

10. N_2O_5

11. NaOH

12. CO_2

13. HF

14. $Pb(OH)_2$

15. NH_4NO_3

16. $NaHCO_3$

17. HgO

18. $Zn(NO_2)_2$

19. H_3PO_4

20. CsOH

21. Li_2O

22. $Ca(OH)_2$

23. $CaBr_2$

24. Fe_2O_3

25. H_2SO_4

26. $FeCO_3$

27. SO_3

28. $Ba(BrO_3)_2$

29. $Al(OH)_3$

30. $HClO_4$

31. $NaC_2H_3O_2$

32. Na_2SO_3

33. H_2CO_3

34. HFO_2

35. NH_4IO_3

36. LiH

(continued)

37. CO

38. $MgBr_2$

39. $SnBr_2$

40. N_2O

41. NH_4F

42. $AsCl_5$

43. $KHCO_3$

44. K_2O

45. Ba_3As_2

46. ZnO

47. $NaClO$

48. SrS

49. $Al(BrO_3)_3$

50. SbF_3

51. $Pd(CN)_2$

52. $ZnSiO_3$

53. $Mg(C_2H_3O_2)_2$

54. $Ca(MnO_4)_2$

55. $Be(NO_3)_2$

56. $NiSeO_4$

57. $RaBr_2$

58. $NaMnO_4$

59. PbI_2

60. CaS

61. Bi_2Te_3

62. $KClO_4$

63. $HgBr_2$

64. $CoSi$

65. P_3N_5

66. $CuSO_3$

67. $FePO_4$

68. $PbTe$

69. $HgNO_3$

70. K_2SiO_3

71. $AgC_2H_3O_2$

72. TeI_4

73. $Zn_3(PO_4)_2$

74. Ag_2S

75. $Cd(HCO_3)_2$

76. ZnF_2

(continued)

77. H_2SO_3

78. $Ba(OH)_2$

79. PbS

80. NaH_2PO_4

81. $NH_4C_2H_3O_2$

82. Ag_3N

83. SiI_4

84. $ZnCO_3$

85. H_3PO_3

86. SnI_4

87. $Pb(NO_3)_2$

88. NaF

89. $KAl(SO_4)_2$

90. KUO_4

91. $SmCl_3$

92. K_2S_5

93. $Fe_3[Fe(CN)_6]_2$

94. $PtCl_2$

95. PtI_4

96. NI_3

97. $MoCl_5$

98. $La(NO_3)_3$

99. Dy_2O_3

100. V_2O_5

Group II. Write the correct formula for each of the following compounds.

1. sulfuric acid

2. sodium hydroxide

3. sodium bromide

4. barium hydroxide

5. calcium oxide

6. hydrosulfuric acid

7. lithium sulfate

8. carbon monoxide

(continued)

Formulas and Nomenclature *(continued)*

9. manganese dioxide

10. sulfur dioxide

11. iron(II) sulfate

12. hypochlorous acid

13. potassium permanganate

14. silver chloride

15. copper(II) hydroxide

16. ammonium sulfide

17. nickel(II) bromide

18. iron(II) oxide

19. bromic acid

20. ammonium bisulfate

21. mercury(I) sulfate

22. iron(III) oxide

23. magnesium phosphate

24. nickel(II) bicarbonate

25. zinc hydroxide

26. hydriodic acid

27. diphosphorous pentoxide

28. aluminum phosphate

29. hydrogen acetate

30. copper(II) nitrate

31. nitrogen dioxide

32. phosphorus trichloride

33. sodium phosphate

34. potassium carbonate

35. phosphoric acid

36. lead(IV) chloride

37. tin(II) bromide

38. ammonium hydroxide

39. periodic acid

40. iron(II) hydroxide

41. carbon dioxide

42. dinitrogen pentoxide

43. silver oxide

44. aluminum nitride

45. manganese(II) hydroxide

46. ammonium carbonate

47. aluminum oxide

48. antimony pentasulfide

(continued)

49. barium carbonate

50. calcium phosphate

51. cesium carbonate

52. potassium silicate

53. silver chromate

54. magnesium sulfite

55. chromium(III) phosphide

56. cobalt(III) nitrate

57. zinc iodide

58. iron(II) fluoride

59. nickel(II) selenide

60. sodium bisulphate

61. lithium oxide

62. copper(I) carbonate

63. strontium carbonate

64. mercury(I) sulfate

65. potassium dichromate

66. manganese(II) oxide

67. nickel(II) chloride

68. lead(II) acetate

69. mercury(II) nitride

70. lead(II) hydroxide

71. tin(IV) chloride

72. selenium tetrafluoride

73. phosphorus pentabromide

74. mercury(I) iodate

75. iron(III) sulfate

76. nickel(II) sulfate

77. silicon dioxide

78. lithium phosphate

79. potassium antimonide

80. nitric acid

81. magnesium nitride

82. cadmium nitrite

83. zinc acetate

84. hydrogen nitrite

85. strontium hydroxide

86. lead(II) sulfate

(continued)

87. aluminum bisulfate

88. disodium hydrogen phosphate

89. ammonium aluminum sulfate

90. copper(II) sulfate pentahydrate

91. lead(II) nitrate

92. gold(III) chloride

93. tin(II) hydroxide

94. hydrogen carbonate

95. ammonium bromate

96. scandium bromide

97. bromine iodide

98. rubidium carbonate

99. potassium thiosulfate

100. potassium arsenate

101. silver potassium cyanide

102. sodium cyanate

103. permanganic acid

104. osmium tetrachloride

105. lanthanum oxide

106. germanium tetrachloride

107. erbium acetate

108. ytterbium oxide

109. calcium hydride

110. iron(II) ferricyanide

Chemistry Problems

Equations

Name _____

Date _____

Write a balanced chemical equation to represent each of the following chemical reactions.

1. iron + sulfur → iron(II) sulfide

2. zinc + copper(II) sulfate → zinc sulfate + copper

3. silver nitrate + sodium bromide → sodium nitrate + silver bromide

4. potassium chlorate (heated) → potassium chloride + oxygen

5. water (electricity) → hydrogen + oxygen

6. mercury(II) oxide (heated) → mercury + oxygen

7. potassium iodide + lead(II) nitrate → lead(II) iodide + potassium nitrate

8. aluminum + oxygen → aluminum oxide

9. magnesium chloride + ammonium nitrate → magnesium nitrate + ammonium chloride

10. iron(III) chloride + ammonium hydroxide → iron(III) hydroxide + ammonium chloride

11. sodium peroxide + water → sodium hydroxide + oxygen

12. iron(III) oxide + carbon → iron + carbon monoxide

13. iron + water → hydrogen + iron(III) oxide

14. iron(III) chloride + potassium hydroxide → potassium chloride + iron(III) hydroxide

15. aluminum + sulfuric acid → aluminum sulfate + hydrogen

16. sodium carbonate + calcium hydroxide → sodium hydroxide + calcium carbonate

(continued)

17. carbon dioxide + water → carbonic acid

18. phosphorus + oxygen → diphosphorus pentoxide

19. sodium + water → sodium hydroxide + hydrogen

20. zinc + sulfuric acid → zinc sulfate + hydrogen

21. aluminum sulfate + calcium hydroxide → aluminum hydroxide + calcium sulfate

22. calcium oxide + water → calcium hydroxide

23. iron + copper(I) nitrate → iron(II) nitrate + copper

24. iron(II) sulfide + hydrochloric acid → hydrogen sulfide + iron(II) chloride

25. potassium oxide + water → potassium hydroxide

26. ammonium sulfide + lead(II) nitrate → ammonium nitrate + lead(II) sulfide

27. mercury(II) hydroxide + phosphoric acid → mercury(II) phosphate + water

28. potassium hydroxide + phosphoric acid → potassium phosphate + water

29. calcium chloride + nitric acid → calcium nitrate + hydrochloric acid

30. potassium carbonate + barium chloride → potassium chloride + barium carbonate

31. magnesium hydroxide + sulfuric acid → magnesium sulfate + water

32. sulfur dioxide + water → sulfurous acid

33. sodium carbonate + hydrochloric acid → sodium chloride + water + carbon dioxide

34. magnesium + nitric acid → magnesium nitrate + hydrogen

(continued)

Chemistry Problems

35. aluminum + iron(III) oxide → aluminum oxide + iron

36. potassium phosphate + magnesium chloride → magnesium phosphate + potassium chloride

37. ammonia + oxygen → nitrogen + water

38. calcium carbonate (heated) → calcium oxide + carbon dioxide

39. sodium chloride + sulfuric acid → sodium sulfate + hydrochloric acid

40. fluorine + sodium hydroxide → sodium fluoride + oxygen + water

41. magnesium nitrate + calcium iodide → calcium nitrate + magnesium iodide

42. aluminum sulfate + ammonium bromide → aluminum bromide + ammonium sulfate

43. potassium fluoride + barium bromide → barium fluoride + potassium bromide

44. copper(II) nitrate + ammonium hydroxide → copper(II) hydroxide + ammonium nitrate

45. sodium nitrate (heated) → sodium nitrite + oxygen

46. lead(II) hydroxide (heated) → lead(II) monoxide + water

47. ammonia + sulfuric acid → ammonium sulfate

48. hydrochloric acid + ammonia → ammonium chloride

49. copper(II) sulfate + iron → iron(II) sulfate + copper

50. aluminum + hydrochloric acid → aluminum chloride + hydrogen

51. carbon + oxygen → carbon dioxide

52. calcium bicarbonate + calcium hydroxide → calcium carbonate + water

(continued)

53. hydrogen sulfide + oxygen → water + sulfur

54. sodium hydroxide + calcium nitrate → sodium nitrate + calcium hydroxide

55. potassium iodide + chlorine → potassium chloride + iodine

56. sulfuric acid + potassium hydroxide → potassium sulfate + water

57. carbon dioxide + carbon → carbon monoxide

58. calcium sulfate + sodium carbonate → calcium carbonate + sodium sulfate

59. water + diphosphorous pentoxide → phosphorus acid

60. aluminum + phosphoric acid → hydrogen + aluminum phosphate

61. ammonium chloride + sodium nitrite → sodium chloride + nitrogen + water

62. chlorine + sodium hydroxide → sodium chloride + sodium hypochlorite + water

63. lead(II) nitrate (heated) → lead monoxide + nitrogen dioxide + oxygen

64. mercury(I) oxide + oxygen → mercury(II) oxide

65. calcium oxide + magnesium chloride → magnesium oxide + calcium chloride

66. calcium + water → calcium hydroxide + hydrogen

67. chromium(III) chloride + sulfuric acid → chromium(III) sulfate + hydrochloric acid

68. iron(III) nitrate + ammonium hydroxide → iron(III) hydroxide + ammonium nitrate

69. aluminum chloride + potassium phosphate → aluminum phosphate + potassium chloride

(continued)

70. aluminum oxide + carbon + chlorine → carbon monoxide + aluminum chloride

71. copper(I) oxide + hydrochloric acid → copper(I) chloride + water

72. magnesium bicarbonate + hydrochloric acid → magnesium chloride + water + carbon dioxide

73. iron + oxygen → iron(III) oxide

74. silicon + water (heated) → silicon dioxide + hydrogen

75. iron(III) oxide + carbon monoxide → iron + carbon dioxide

76. calcium chloride + chromium(III) nitrate → calcium nitrate + chromium(III) chloride

77. zinc sulfide + oxygen → zinc oxide + sulfur dioxide

78. calcium phosphate + sulfuric acid → calcium sulfate + phosphoric acid

79. iron(III) hydroxide (heated) → iron(III) oxide + water

80. aluminum sulfate + sodium bicarbonate → aluminum hydroxide + sodium sulfate + carbon dioxide

81. calcium phosphate + silicon dioxide + carbon → phosphorus + calcium silicate + carbon monoxide

82. calcium oxide + sulfur dioxide → calcium sulfite

83. carbon dioxide + magnesium hydroxide → magnesium carbonate + water

84. calcium oxide + hydrochloric acid → calcium chloride + water

85. calcium carbonate + silicon dioxide → calcium silicate + carbon dioxide

86. antimony + chlorine → antimony trichloride

(continued)

Chemistry Problems

87. magnesium nitride + water → magnesium hydroxide + ammonia

88. arsenic + oxygen → arsenic(III) oxide

89. ammonium bicarbonate (heated) → ammonia + water + carbon dioxide

90. copper(II) oxide + ammonia → copper + water + nitrogen

91. ammonium dichromate (heated) → chromium trioxide + nitrogen + water

92. hydrogen sulfide + cadmium nitrate → nitric acid + cadmium sulfide

93. barium bromide + sodium phosphate → barium phosphate + sodium bromide

94. aluminum chloride + ammonium fluoride → aluminum fluoride + ammonium chloride

95. silver nitrate + potassium sulfate → silver sulfate + potassium nitrate

96. bismuth nitrate + calcium iodide → bismuth iodide + calcium nitrate

97. aluminum chromate + ammonium sulfate → ammonium chromate + aluminum sulfate

98. zinc nitrate + ammonium bromide → zinc bromide + ammonium nitrate

99. bismuth nitrate + ammonium hydroxide → bismuth hydroxide + ammonium nitrate

100. cadmium nitrate + sulfuric acid → cadmium sulfate + nitric acid

101. zinc + silver iodide → zinc iodide + silver

102. iron(III) chloride + sulfuric acid → iron(III) sulfate + hydrochloric acid

103. bismuth sulfate + ammonium hydroxide → bismuth hydroxide + ammonium sulfate

104. hydrogen iodide + oxygen → iodine + water

(continued)

105. potassium sulfate + barium chloride → barium sulfate + potassium chloride

106. barium sulfate + carbon → barium sulfide + carbon monoxide

107. aluminum oxide + hydrofluoric acid → aluminum fluoride + water

108. aluminum fluoride + sulfuric acid → aluminum sulfate + hydrogen fluoride

109. potassium iodide + hydrogen peroxide → potassium hydroxide + iodine

110. zinc + iron(III) sulfate → zinc sulfate + iron(II) sulfate

111. lead(II) sulfide + lead monoxide → lead + sulfur dioxide

112. copper + sulfuric acid → copper(II) sulfate + water + sulfur dioxide

113. aluminum hydroxide (heated) → aluminum oxide + water

114. nitrogen + hydrogen → ammonia

115. sodium carbonate + carbonic acid → sodium bicarbonate

116. silicon dioxide + hydrofluoric acid → water + silicon tetrafluoride

117. sodium hypochlorite → sodium chloride + sodium chlorate

118. sodium chlorite + chlorine → sodium chloride + chlorine dioxide

119. methane + sulfur dioxide (heated) → hydrogen sulfide + carbon dioxide + hydrogen

120. tellurous acid → tellurium dioxide + water

121. iron(II) selenide + hydrochloric acid → iron(II) chloride + hydrogen selenide

122. magnesium + nitrogen → magnesium nitride

(continued)

123. silver cyanide + potassium → potassium cyanide + silver

124. copper(II) sulfate + ammonia → cupriammonia sulfate

125. calcium carbide + nitrogen → calcium cyanamide + carbon

126. calcium cyanamide + water → calcium carbonate + ammonia

127. zinc arsenide + hydrochloric acid → arsine + zinc chloride

128. lead(II) hydroxide + sodium stannite → lead + sodium stannate + water

129. sodium silicate + hydrochloric acid → sodium chloride + silicic acid

130. boron trioxide + magnesium → magnesium oxide + boron

131. iron(II) cyanide + potassium cyanide → potassium ferrocyanide

132. sodium aluminate + ammonium chloride → ammonium aluminum oxide + sodium chloride

133. aluminum hydroxide + sodium hydroxide → sodium aluminate + water

134. tungsten + chlorine (heated) → tungsten hexachloride

135. calcium + ammonia → calcium hydride + nitrogen

136. sodium wolframate + sulfuric acid → sodium sulfate + tungstic acid

137. lithium hydride + water → lithium hydroxide + hydrogen

138. boric acid → tetraboric acid + water

139. zinc hydroxide + potassium hydroxide → potassium zincate + water

140. nickel + carbon monoxide → nickel carbonyl

Reaction Prediction

Name _____

Date _____

In each of the following examples:

(a) State what type of reaction is expected.

(b) Tell whether the reaction will occur or not, and why.

(c) Write the balanced equation for those reactions that do take place; write the symbols and formulas of the reactants for those reactions that do occur.

(d) Indicate whether double replacement reactions are reversible or irreversible.

1. aluminum plus hydrochloric acid

2. calcium hydroxide plus nitric acid

3. aluminum plus magnesium

4. magnesium plus zinc nitrate

5. mercury plus oxygen

6. zinc chloride plus hydrogen sulfide

7. dinitrogen pentoxide plus water

8. silver chloride plus sodium nitrate

9. sodium chlorate (heated)

10. barium nitrate plus sodium chromate

11. sodium bromide plus silver nitrate

12. calcium phosphate plus aluminum sulfate

13. zinc carbonate (heated)

14. mercury(I) sulfate plus ammonium nitrate

(continued)

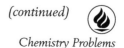

15. potassium plus fluorine

16. potassium nitrate plus zinc phosphate

17. lithium oxide plus water

18. sodium chloride (electrolyzed)

19. silver plus barium

20. iron(III) hydroxide plus phosphoric acid

21. sodium plus nitric acid

22. iron(III) iodide plus cupric nitrate

23. copper plus sulfuric acid

24. lead plus potassium chlorate

25. sulfur dioxide plus water

26. oxygen plus sulfur

27. sodium sulfate plus barium chloride

28. ammonium phosphate plus lithium hydroxide

29. hydrogen plus oxygen

30. mercury plus nitric acid

31. sodium oxide plus water

32. calcium carbonate plus lithium chloride

(continued)

33. mercury(I) sulfate plus hydrochloric acid

34. potassium nitrate (heated)

35. chlorine plus bromine

36. mercury(I) nitrate plus sodium carbonate

37. magnesium plus hydrochloric acid

38. water (electrolyzed)

39. ammonium nitrite plus barium hydroxide

40. ammonium sulfate plus calcium hydroxide

41. mercury(II) oxide (heated)

42. ammonium phosphate plus aluminum chloride

43. barium oxide plus water

44. iron(III) hydroxide plus nitric acid

45. calcium plus oxygen

46. calcium plus phosphoric acid

47. calcium chloride plus ammonium hydroxide

48. aluminum sulfide plus hydrochloric acid

49. magnesium plus sulfur

50. calcium plus aluminum chloride

(continued)

Chemistry Problems

51. potassium hydroxide plus hydrosulfuric acid

52. sodium carbonate plus sulfuric acid

53. barium sulfate plus calcium chloride

54. silver plus mercury(I) nitrate

55. barium carbonate (heated)

56. lithium plus bromine

57. sodium chloride plus potassium chromate

58. potassium sulfide plus iron(II) nitrate

59. iodine plus ammonium fluoride

60. sodium plus calcium

61. aluminum chloride (electrolyzed)

62. lead(II) chlorate plus sodium sulfide

63. sulfur trioxide plus water

64. calcium carbonate plus hydrochloric acid

65. iron plus sodium bromide

66. ammonium acetate plus iron(II) chloride

67. silver bromide plus ammonium sulfate

68. zinc plus sulfuric acid

(*continued*)

69. neon plus potassium

70. iron plus potassium iodide

71. lead(II) hydroxide plus hydrochloric acid

72. iron plus sulfur

73. potassium chlorate (heated)

74. oxygen plus chlorine

75. silver iodide plus iron(III) sulfide

76. iron(II) carbonate plus phosphoric acid

77. potassium iodide plus ammonium nitrate

78. potassium plus sodium nitrate (heated)

79. bromine plus sodium chloride

80. silver sulfide plus hydrochloric acid

81. magnesium nitrate plus hydrochloric acid

82. ammonia plus hydrogen chloride

83. zinc hydroxide plus sulfuric acid

84. calcium oxide plus water

85. sodium plus chlorine

86. calcium hydroxide (heated)

(continued)

87. fluorine plus potassium bromide

88. ammonium hydroxide plus sulfuric acid

89. sodium chloride plus potassium nitrate

90. lead plus tin(II) nitrate

91. carbon dioxide plus water

92. chlorine plus lithium bromide

93. lithium hydroxide plus phosphoric acid

94. potassium sulfite plus nitric acid

95. ammonium chloride plus potassium hydroxide

96. strontium carbonate plus nitric acid

97. tin plus mercury(I) nitrate

98. ammonium sulfite plus hydrochloric acid

99. magnesium carbonate plus phosphoric acid

100. aluminum sulfite plus hydrochloric acid

Density and Specific Gravity

Name _____

Date _____

Solids and Liquids

1. Copper has a density of 8.9 g/cm^3. What is its specific gravity?

2. 8.0 cm^3 of platinum are found to have a mass of 171.2 g. Determine the density and specific gravity of platinum.

3. 50.0 cm^3 of lead have a mass of 565 g. What is the density of lead? The specific gravity?

4. Chloroform has a specific gravity of 1.5. What is its density? What is the mass of 1.0 dm^3 of chloroform?

5. 127 cm^3 of turpentine are found to have a mass of 114.3 g. What is the specific gravity of turpentine?

6. A block of granite with a volume of 3.8 m^3 has a mass of 638.30 kg. What is its specific gravity?

7. What is the mass of 137 cm^3 of nickel? The specific gravity of nickel is 8.8.

8. Calculate the specific gravity of marble from the fact that 321 cm^3 of the material have a mass of 866.7 g.

9. Steel has a specific gravity of 7.8. What volume of steel would have a mass of 1.00 kg?

10. What is the mass of 1.2 dm^3 of pine wood? The specific gravity of pine wood is 0.4.

11. 78 cm^3 of soft coal have a mass of 101.4 g. What is its specific gravity?

12. Brass has a specific gravity of 8.5. What would be the mass of 1.0 m^3 of brass in kg?

13. The concentrated sulfuric acid in a 500.0 cm^3 flask has a mass of 920 g. What is the specific gravity of sulfuric acid?

(continued)

14. 421 cm^3 of water at 4 °C are cooled to 0 °C. The new volume is 425 cm^3. What is the density of water at 0 °C?

15. Cast iron has a specific gravity of 7.1. What is the density of cast iron in kilograms per cubic meter?

16. What is the mass of 62.3 cm^3 of carbon tetrachloride if the specific gravity of carbon tetrachloride is 1.6?

17. Carbon disulfide may be purchased in bottles containing 100.0 kg of the disulfide. These bottles have a volume of 1.27 m^3. What is the specific gravity of carbon disulfide? What is the weight of 1.00 dm^3 of carbon disulfide?

18. 50.0 mL of water at 4 °C are frozen to ice. The volume of the ice produced is 54.5 cm^3. What is the density and specific gravity of the ice?

19. Seawater has a specific gravity of 1.03. A certain sample of seawater was found to have a mass of 154 g. What was the volume of the water?

20. What is the mass of 5.0 dm^3 of mercury? The specific gravity of mercury is 13.6.

Gases and Vapors

Note: In the following problems, **air** is the standard used in determining specific gravity.

1. 500.0 cm^3 of hydrogen chloride have a mass of 0.819 6 g. Find the density and specific gravity of this gas.

2. 1.0 dm^3 of ammonia has a mass of 0.771 0 g. Find the density and specific gravity of ammonia.

3. The density of nitrous oxide is 1.977 8 g/dm^3. Calculate the molecular mass and specific gravity of this gas.

4. The specific gravity of carbon monoxide is 0.967. What is its density and molecular mass?

(continued)

5. The specific gravity of chlorine is 2.49. What is its density and molecular mass?

6. The density of hydrogen is 0.089 8 g/dm^3. What is its specific gravity and molecular mass?

7. 250 cm^3 of krypton gas have a mass of 0.934 g. Find the density, specific gravity, and molecular mass of kyrpton.

8. Find the mass of 5.0 dm^3 of carbon dioxide.

9. The specific gravity of stibine is 4.34. Find the density and molecular mass of this gas.

10. 100.0 cm^3 of silicon tetrafluoride have a mass of 0.461 g. Find the density, specific gravity, and molecular mass of the gas.

11. Find the density of chlorine.

12. The density of phosphorus trifluoride is 3.90 g/dm^3. Find its specific gravity and molecular mass.

13. Find the mass of 200.0 cm^3 of sulfur monochloride, S_2Cl_2.

14. Determine the specific gravity of methane, CH_4.

15. 400.0 cm^3 of hydrogen iodide have a mass of 2.28 g. Find the density, specific gravity, and molecular mass.

16. What is the specific gravity of bromine vapor?

17. Find the mass of 100.0 cm^3 of xenon.

18. The specific gravity of carbonyl sulfide, COS, is 2.10. Find its density and molecular mass.

(continued)

Name _____

Date _____

19. The density of an unknown gas is 2.09 g/dm^3. Find its specific gravity and molecular mass.

20. 300.0 cm^3 of sulfur dioxide have a mass of 0.878 1 g. Find the density, specific gravity, and molecular mass of this gas.

21. The density of helium if 0.178 5 g/dm^3. Find the specific gravity and molecular mass of helium.

22. Find the mass of 450.0 cm^3 of hydrogen sulfide.

23. The specific gravity of nitric oxide is 1.037. Find its density and molecular mass.

24. What is the specific gravity of argon?

25. 50.0 cm^3 of oxygen weigh 0.071 4 g. Find the density, specific gravity, and molecular mass of oxygen.

26. What is the mass of 285 cm^3 of tellurium hydride, TeH_2?

27. Find the density and specific gravity of nitrogen.

28. The specific gravity of hydrogen fluoride is 0.988. What molecular mass does this give for the compound?

29. What is the mass of 5.2 dm^3 of phosgene, $COCl_2$?

30. Calculate the density and specific gravity of hydrogen bromide.

Gas Law Problems

Name _____

Date _____

Group I. Correct the volumes of the dry gases as directed in each of the following problems.

1. Change 125 mL of a gas at 25 °C to standard temperature.

2. Change 300.0 cm^3 of a gas at 0.0 °C to 30.0 °C.

3. Change 220.0 mL of a gas at 10.0 °C to 100.0 °C.

4. Change 1.00 L of a gas at 32 °C to 27 °C.

5. Change 100.0 cm^3 of a gas at 98.8 kPa pressure to standard pressure.

6. Change 250.0 mL of a gas at standard pressure to 104 kPa.

7. Change 30.0 cm^3 of a gas at standard pressure to 80.0 kPa.

8. Change 750.0 mL of a gas at 93.3 kPa to 106.6 kPa.

9. Change 500.0 cm^3 of a gas at 60.0 °C and 106.6 kPa to standard conditions.

10. Change 800.0 mL at 40.0 °C and 700.0 mmHg to standard conditions.

11. Change 60.0 cm^3 at standard conditions to 55 °C and 99.3 kPa.

12. Change 35 mL at standard conditions to 25 °C and 96.6 kPa.

13. Change 50.0 cm^3 at standard conditions to 43 °C and 750 mmHg.

14. Change 75.0 mL at 100.0 °C and 60.0 kPa to standard conditions.

15. Change 10.0 cm^3 at 27 °C and 126.6 kPa atm to standard conditions.

16. Change 250 mL at 32 °C and 100 kPa to 47 °C and 104 kPa.

17. Change 200.0 cm^3 at 17 °C and 800.0 mmHg to 37 °C and 700.0 mm.

18. Change 45 mL at 15 °C and 105.3 kPa to 23 °C and 108 kPa.

19. Change 135 cm^3 at 34 °C and 106.6 kPa to 20 °C and 101.3 kPa.

20. Change 310 mL at 58 °C and 97.3 kPa to 15 °C and 100.0 kPa.

21. Change 2.0 dm^3 at 43 °C and 96.0 kPa to 28 °C and 106.6 kPa.

22. Change 550 cm^3 at 25 °C and 790 mmHg to 55 °C and 700.00 mmHg.

(continued)

Chemistry Problems

23. Change 20.0 mL at 33 °C and 93.3 kPa to 57 °C and 109.4 kPa.

24. Change 150 cm^3 at 0.0 °C and 98.6 kPa to 41 °C and 100.0 kPa.

25. Change 375 mL at 13 °C and 104.0 kPa to 33 °C and 101.3 kPa.

26. Change 80.0 cm^3 at 27 °C and 150 mmHg to standard conditions.

27. Change 37.5 mL at –5 °C and 600.0 mmHg to standard conditions.

28. Change 175 cm^3 at 3.0 °C and 96.0 kPa to standard conditions.

29. Change 350 mL at –30. °C and 100.0 mmHg to standard conditions.

30. Change 2.0 L at –45 °C and 25 mmHg to standard conditions.

31. Change 18 cm^3 at standard conditions to 23 °C and 98.6 kPa.

32. Change 120. mL at standard conditions to 60.0 °C and 850 mmHg.

33. Change 36 cm^3 at standard conditions to –13 °C and 320 mmHg.

34. Change 3.2 dm^3 at standard conditions to 110 °C and 720 mmHg.

35. Change 42 mL at standard conditions to 27 °C and 105.8 kPa.

36. Change 272 cm^3 at standard conditions to 31 °C and 324.2 kPa.

37. Change 4.0 L at standard conditions to –43 °C and 420 mmHg.

38. Change 75 dm^3 at standard conditions to 17 °C and 98.0 kPa.

39. Change 360 cm^3 at standard conditions to –19 °C and 48.0 kPa.

40. Change 6.5 L at standard conditions to –23 °C and 150 mmHg.

Group II. What pressure is needed to make the following changes?

1. 130 mL of a dry gas at 98.6 kPa to 150 mL

2. 25 cm^3 of a dry gas at 65 mmHg to 30.0 mL

3. 1.0 dm^3 of a dry gas at 70.0 mmHg to 1.2 dm^3

4. 75 mL of a dry gas at 415 kPa and 27 °C to 70.0 mL at 25 °C

5. 60.0 cm^3 of a dry gas at 101.3 kPa and 0.0 °C to 10.0 cm^3 at 25 °C

6. 400.0 mL of a dry gas at 760 mmHg and 15 °C to 300.0 mL at –30.0 °C

(continued)

Group III. What temperature is needed to make the following changes?

1. 30.0 cm^3 of a dry gas at 14 °C to 22 cm^3

2. 16.4 mL of a dry gas at 28 °C to 20.0 mL

3. 39 cm^3 of a dry gas at 0.0 °C to 35 cm^3

4. 50.0 mL of a dry gas at 5 °C and 760 mmHg to 55 mL and 780 mmHg

5. 1.0 L of a dry gas at 10.0 °C and 106.6 kPa to 0.50 L and 101.3 kPa

6. 10.0 cm^3 of a dry gas at 20.0 °C and 101.3 kPa to 1.0 cm^3 and 106.6 kPa

Group IV. Solve the following gas law problems.

1. If 120 mL of oxygen are collected over water at 27 °C and 740 mmHg pressure, what will the volume of the dry gas be at STP?

2. If 500.0 mL of hydrogen are collected over water at 20.0 °C and 99.3 kPa, what will the volume of the dry gas be at STP?

3. If 250 mL of nitrogen are collected over water at 25 °C and 100.0 kPa, what will the volume of the dry gas be at STP?

4. A certain gas is collected over water at 740 mmHg and 23 °C. The collecting tube is left in place, and the volume is not measured until the next day when the pressure is 745 mm and the temperature 20.0 °C, at which time the volume is found to be 15.3 mL. What was the original volume?

5. If 153.48 mL of carbon dioxide are collected over mercury at 14 °C and 96.4 kPa, what will the volume be at STP?

6. 113 mL of oxygen are collected over water at 22 °C and 98.8 kPa and left in position overnight. On the next day, the volume has reduced to 109 mL and the temperature reads 21 °C. What is the pressure on the second day?

7. If 18 mL of oxygen are collected over water at 27 °C and 740 mmHg of pressure, what volume will the dry gas have at STP?

(continued)

Chemistry Problems

8. 36 mL of nitrogen are collected over water at 25 °C. The barometer is broken and no pressure can be read. Three days later, when a new barometer arrives, the volume of the damp gas has changed to 32 mL at a temperature of 21 °C. The barometric reading is 98.5 kPa. What was the original pressure?

Group V. Listed below are the densities for a number of gases at standard conditions. Find the new density of the gas in each case at the stated conditions.

1. carbon dioxide: 1.98 g/dm³ (at 40.0 °C)

2. helium: 0.178 g/dm³ (at 23 °C)

3. nitric oxide: 1.34 g/dm³ (at 750 mmHg)

4. hydrogen chloride: 1.64 g/dm³ (at 104 kPa)

5. acetylene: 1.17 g/dm³ (at 37 °C and 104 kPa)

6. oxygen: 1.43 g/dm³ (at 17 °C and 740 mmHg)

7. sulfur dioxide: 2.93 g/dm³ (at 25 °C and 100.6 kPa)

8. ethylene: 1.26 g/dm³ (at 15 °C and 700 mmHg)

9. chlorine: 3.21 g/dm³ (at 8.0 °C and 795 mmHg)

10. nitrogen: 1.25 g/dm³ (at –23 °C and 53.3 kPa)

11. ammonia: 0.77 g/dm³ (at 500.0 °C and 250 mmHg)

12. methane: 0.717 g/dm³ (at 250 °C and 602.5 kPa)

13. hydrogen sulfide: 1.54 g/dm³ (at –53 °C and 2550 mmHg)

(continued)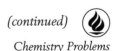

14. hydrogen: 0.09 g/L (at 2850 °C and 1.013 kPa)

15. argon: 1.78 g/L (at –150 °C and 22,800 kPa)

16. 25 mL of fluorine gas have a mass of 0.0390 g at 25 °C. What is the density of fluorine at standard conditions?

17. Find the density of carbon monoxide gas at standard conditions if $30\overline{0}$ mL of the gas have a mass of 0.395 g at 800.0 mmHg of pressure.

18. 10.0 mL of neon would have a mass of 10.8 g at 1333 kPa and –270 °C. From this information, calculate the density of neon at standard conditions.

19. The mass of 230 mL of hydrogen at 506.5 kPa and –53 °C is 0.128 g. What is the density of hydrogen at standard conditions?

20. The mass of 15 mL of nitrous oxide at 50.0 mmHg and 2000.0 °C is 2.3×10^{-4} g. What would be the mass of 1.0 L of this gas at standard conditions?

21. A student collected a liter flask full of oxygen over water when the thermometer read 23 °C and the barometer read 750 mmHg. Two hours later, the volume of the gas in the flask had reduced to 978 mL although the barometer had not changed. What change had occurred in the temperature?

22. A large balloon containing 400.0 L of hydrogen is released from the earth when the temperature is 21 °C and the pressure 760 mmHg. What volume will the gas occupy when it reaches the atmospheric level at which the temperature is –67 °C and the pressure 16.7 kPa?

23. Suppose the balloon in problem 22 reached the given level, but the temperature was +67 °C instead of –67 °C. (All other conditions as stated in 22.) What volume would the hydrogen occupy then?

24. A quantity of nitrogen gas is enclosed in a tightly stoppered 500.0 mL flask at room temperature (20.0 °C) and 760 mmHg pressure. The flask is then heated to 680 °C. If the flask can withstand pressures of less than 303.9 kPa, will it explode under this heating?

(continued)

Name _____

Date _____

25. A 1.0 L rubber bladder is filled with carbon dioxide gas in a warm (25 °C) room (pressure = 99.3 kPa). What volume will the gas occupy when it is taken out into the open air where the temperature is –12 °C and the pressure, 98.9 kPa?

26. 500.0 mL of air are trapped in a tube over mercury at 25 °C. It is found that, after six days, the air has expanded so that 32 mL have escaped from the tube. What total temperature change occurred over this period if the pressure remained constant?

27. A rubber balloon containing 1.0 L of gas is carried from the top of a mountain to the bottom of the mountain, where its volume is measured as 0.85 L at standard pressure. Assuming that there was no temperature change during the trip, what was the pressure at the top of the mountain?

28. A sudden cold snap in June causes Mr. Van Dellen to order a 15 L tank of butane gas for his cottage heater. The temperature on this day was 18 °C and the atmospheric pressure, 743 mmHg. Assuming that some type of piston arrangement was built into the tank so that the gas could expand freely, what volume would the butane gas occupy a week later when the temperature reached 42 °C and 721 mmHg of pressure (assume 3.0 L of gas had been used)?

29. A toy balloon containing 425 mL of air escapes from a little boy watching a parade. The temperature is 32 °C at street level and the pressure is 99.3 kPa. When the balloon stops rising, its volume has become 895 mL although the atmospheric pressure has decreased by only 300.0 torr. What is the temperature at this level?

30. To what temperature must 15 L of oxygen gas at 0.0 °C be heated at 101.3 kPa pressure in order to occupy a volume of 23 L, assuming that the pressure increases by 5.0 mmHg?

Group VI. Solve the following problems involving the ideal gas law equation.

1. What pressure is exerted by 1.0 mol of an ideal gas contained in a 1.0 dm^3 vessel at 0.0 °C?

2. What volume will 5.0 mol of an ideal gas occupy at 25.0 °C and 152 kPa of pressure?

3. Calculate the molecular mass of a gas if 4.5 L of the gas at 104.6 kPa and 23.5 °C have a mass of 13.5 g.

(continued)

4. 0.453 mol of a gas confined to a 15.0 L container exerts a pressure of 125.6 kPa on the walls of the container. What is the temperature of the gas?

5. 5.4 g of carbon dioxide gas are confined to a 20.0 L container at a temperature of 32.5 °C. What pressure does the gas exert?

6. 2.125 g of a gas in a 1.25 L container exert a pressure of 84.9 kPa at 40.0 °C. What is the molecular mass of the gas?

7. To what temperature must 10.0 g of ammonia gas be heated in a 15.0 L container in order for it to exert a pressure of 354.5 kPa?

8. 2.0×10^{-5} g of hydrogen gas at 155 °C exerts a pressure of 43.0 kPa on the walls of a small cylindrical tube. What is the volume of the tube?

Group VII. **Listed below are various combinations of (1) molar composition of a gas mixture (m_A, m_B, m_C, etc.), (2) partial pressure of each gas in the mixture (p_A, p_B, p_C, etc.), and (3) the total pressure of the mixture (p_T). From this information, calculate any of the variables which are not given.**

1. $p_T = 785$ mmHg; $m_A = 45\%$; $m_B = 55\%$

2. $p_T = 1.050$ kPa; $m_A = 38\%$; $m_B = 40\%$; $m_C = 22\%$

3. $p_A = 420$ kPa; $p_B = 180$ kPa; $p_C = 200$ kPa

4. $p_T = 1.12$ kPa; $p_A = 0.43$ kPa (two gases only)

5. $m_A = 21\%$; $m_B = 34\%$; $m_C = 45\%$; $p_A = 183$ kPa

6. 10.0 g of carbon dioxide; 15.0 g of nitrogen; $p_T = 1.142$ atm

7. 5.0 g of argon; 5.0 g of xenon; 5.0 g of krypton; $p_{Ar} = 86.4$ kPa

8. $p_T = 1.28$ atm; 3.4×10^{-3} g sulfur dioxide; 1.8×10^{-3} g carbon dioxide; 4.0×10^{-2} g nitrogen

(continued)

Group VIII. Find the relative rate of diffusion between each of the following pairs of gases.

1. hydrogen and nitrogen

2. oxygen and carbon dioxide

3. hydrogen and carbon dioxide

4. nitrogen and oxygen

5. hydrogen chloride and chlorine

6. carbon monoxide and carbon dioxide

Group IX. Listed below are (1) the relative rate of diffusion between two gases and (2) the density of the less dense gas. From this information, calculate the density of the more dense gas.

1. 1.14; 1.93 g/L

2. 2.04; 1.28 g/L

3. 1.42; 1.13 g/L

4. 1.85; 1.62 g/L

5. 1.63; 1.46 g/L

6. 1.55; 1.39 g/L

Molecular Mass and Mole Calculations

Name _____

Date _____

Group I. Find the molecular mass or formula mass for each of the compounds shown below.

1. H_3PO_4

2. $AlCl_3$

3. $Dy(OH)_3$

4. $K_2C_4H_4O_6$

5. H_2SO_4

6. N_2O_5

7. $CuSO_4 \cdot 5H_2O$

8. $NiSO_4$

9. $Sn(OH)_4$

10. $(NH_4)_3PO_4$

11. $Fe(C_2H_3O_2)_3$

12. SO_2

13. $KAl(SO_4)_2 \cdot 12H_2O$

14. $NaIO_4$

15. $Pr(OH)_3$

16. $K_4Fe(CN)_6$

17. Nd_2O_3

18. $Sb(NO_3)_3$

19. K_3PO_4

20. $Ga_2(SO_4)_3$

21. zinc acetate

22. copper(I) sulfate

23. carbon dioxide

24. calcium bicarbonate

25. nitric acid

26. aluminum nitrate

27. ammonium sulfate

28. barium chloride dihydrate

29. iron(II) phosphate

30. strontium hydroxide

31. sodium sulfite

32. magnesium nitride

(continued)

Molecular Mass and Mole Calculations

33. hydrochloric acid

34. iron(II) sulfate

35. lead(II) bromide

36. cesium chloride

37. copper(II) nitrate

38. magnesium bromide

39. iron(III) hydroxide

40. ammonium carbonate

Group II. Express each of the following in grams.

1. 1.0 mol of KBr

2. 1.0 mol of $CaCl_2$

3. 2.0 mol of AlF_3

4. 0.50 mol of KNO_3

5. 0.30 mol of $NaHCO_3$

6. 0.25 mol of potassium iodate

7. 0.10 mol of iron(III) carbonate

8. 0.20 mol of sodium sulfate decahydrate

9. 8.0 mol of sodium dihydrogen phosphate

10. 0.30 mol of phosphorus pentachloride

11. 1.50 mol of HCl

12. 0.736 mol of H_2SO_4

13. 0.042 mol of KOH

14. 0.147 mol of NH_4OH

(continued)

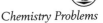

15. 1.26×10^{-4} mol of $HC_2H_3O_2$

16. 2.6 mol of lithium bromide

17. 0.395 mol of magnesium nitrate

18. 2.4×10^{-3} mol of sodium phosphate

19. 4.6×10^{-3} mol of ammonium chloride

20. 3.85×10^{-6} mol of lead(II) ions

Group III. Convert each of the following to its equivalent in moles.

1. 86.84 g of LiBr

2. 62.3 g of MgF_2

3. 17.1 g of H_2S

4. 302.7 g of $ScCl_3$

5. 8.5 g of PH_3

6. 33.4 g of KNO_3

7. 17.2 g of H_3PO_4

8. 38.1 g of $FeCl_3$

9. 15.9 g of $Ca(NO_3)_2$

10. 8.8 g of potassium carbonate

11. 5.0 g of ammonium sulfate

12. 2.5 g of copper(II) chloride

(continued)

Molecular Mass and Mole Calculations

(continued)

13. 30.0 g of tin(II) nitrate

14. 0.257 g of arsenic pentachloride

15. 48.8 g of lanthanum nitrate

16. 0.2000 g of cadmium bromide

17. 1.50×10^{-4} g of lanthanum nitrate

18. 0.0350 g of potassium permanganate

19. 10.00 g of ammonium dichromate

20. 8.045 g of titanium tetrachloride

Percentage Composition

Name _____

Date _____

Find the percentage composition of each of the following compounds.

1. $NaCl$

2. HBr

3. KI

4. CO

5. SO_2

6. H_2Te

7. K_2S

8. AlI_3

9. NH_3

10. NH_4Br

11. $NaNO_3$

12. H_2SO_4

13. $Ca(NO_3)_2$

14. $Sc(OH)_3$

15. K_3PO_4

16. $Zn_3(PO_4)_2$

17. Fe_3O_4

18. $(NH_4)_2SO_4$

19. $CuSO_4 \cdot 5H_2O$

20. $(NH_4)_2Cr_2O_7$

21. potassium bromide

22. cesium chloride

23. nitric oxide

24. hydriodic acid

25. carbon dioxide

26. water

27. sodium oxide

28. phosphorus tribromide

29. arsine

30. potassium nitrite

31. ammonium chloride

32. phosphoric acid

33. barium hydroxide

34. aluminum nitrate

(continued)

Percentage Composition (continued)

Name _____

Date _____

35. lithium phosphate

36. calcium phosphate

37. aluminum oxide

38. ammonium carbonate

39. ammonium phosphate

40. barium chloride dihydrate

41. What is the percentage of strontium in $SrCl_2$?

42. What is the percentage of carbon in calcium carbonate?

43. What is the percentage of water in copper(II) sulfate pentahydrate?

44. What is the percentage of hydrogen in $Ca(C_2H_3O_2)_2$?

45. What is the percentage of oxygen in $KMnO_4$?

46. What is the percentage of cobalt in cobalt(II) nitrate?

47. What is the percentage of carbon in sucrose, $C_{12}H_{22}O_{11}$?

48. What is the percentage of bismuth in sodium bismuthate, $NaBiO_3$?

49. What is the percentage of aluminum in $NaAl(SO_4)_2 \cdot 12H_2O$?

50. What is the percentage of chromium in potassium dichromate?

51. How much iron can be obtained from 100.0 g of Fe_3O_4?

52. How much phosphorus is there in 500.0 g of calcium phosphate?

53. How many kilograms of aluminum can be obtained from 2000 kg of aluminum chloride?

54. How many grams of hydrogen can be obtained from 35 cm^3 of water?

55. How much anhydrous copper(II) sulfate can be obtained from 15.0 g of hydrated copper(II) sulfate?

56. How many grams of silver can be recovered from 10.0 g of silver sulfide?

(continued)

Name _____

Date _____

57. How much lead can be extracted from 200.0 kg of its ore, lead sulfide?

58. Free hydrogen gas can be obtained from ethyl alcohol, C_2H_5OH. How many grams of hydrogen would be obtained from 150.0 g of the alcohol?

59. How many grams of water are driven off by the heating of 125 g of cobalt(II) chloride hexahydrate?

60. Can 18 g of oxygen be obtained by heating 50.0 g of potassium chlorate?

Empirical Formula

Name _____

Date _____

Find the empirical formula for each of the following substances. The percentage composition is given.

1. 88.8% copper; 11.2% oxygen

2. 40.0% carbon; 6.7% hydrogen; 53.3% oxygen

3. 92.3% carbon; 7.7% hydrogen

4. 70.0% iron; 30.0% oxygen

5. 5.88% hydrogen; 94.12% oxygen

6. 79.90% copper; 20.10% oxygen

7. 56.4% potassium; 8.7% carbon; 34.9% oxygen

8. 10.04% carbon; 0.84% hydrogen; 89.12% chlorine

9. 42.50% chromium; 57.50% chlorine

10. 15.8% carbon; 84.2% sulfur

11. 30.43% nitrogen; 69.57% oxygen

12. 82.40% nitrogen; 17.60% hydrogen

13. 12.5% hydrogen; 37.5% carbon; 50.0% oxygen

14. 75.0% carbon; 25.0% hydrogen

15. 29.40% calcium; 23.56% sulfur; 47.04% oxygen

16. 38.67% potassium; 13.85% nitrogen; 47.48% oxygen

17. 60.0% magnesium; 40.0% oxygen

18. 52.94% aluminum; 47.06% oxygen

19. 72.40% iron; 27.60% oxygen

(continued)

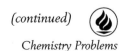

20. 52.0% zinc; 9.6% carbon; 38.4% oxygen

21. 60.98% arsenic; 39.02% sulfur

22. 74.17% mercury; 25.83% chlorine

23. 60.0% terbium; 40.0% chlorine

24. 65.1% scandium; 34.9% oxygen

25. 34.6% gallium; 17.8% carbon; 47.6% oxygen

26. A sample of potassium sulfate has the following composition: 17.96 g of potassium, 7.35 g of sulfur, 14.70 g of oxygen. What is its simplest formula?

27. A pure sample of mercury oxide produced 20.3 g of mercury and 1.7 g of oxygen. What oxide of mercury is this?

28. 11.00 g of a certain compound contain 2.82 g of magnesium of 8.18 g of chlorine. What is its simplest formula?

29. A certain sample of a barium salt contains 8.57 g of barium and 4.43 g of chlorine. What is its simplest formula?

30. 50.0 g of sulfur are mixed with 100.0 g of iron and the mixture is heated. When the reaction is completed, 12.7 g of iron remain. What is the formula of the compound formed?

31. 0.916 g of iron is heated in air. The resulting product has a mass of 1.178 g. What is the formula of the compound?

32. 21.42 g of calcium combine with 8.58 g of oxygen in a combustion reaction. What is the formula of the product?

33. 0.238 g of carbon is burned in 1.00 dm^3 of oxygen. The oxide of carbon which is formed has a mass of 0.872 g. What is the formula of this oxide?

34. Analysis of 100.0 g of a compound produces the following result: 26.6 g of potassium; 35.4 g of chromium; 38.0 g of oxygen. What is the formula of this compound?

35. When 0.982 g of mercuric oxide are heated until all oxygen is expelled, 0.909 g of mercury remain. What is the formula of this oxide?

Molecular Formula

Group I. Given below are the percentage compositions for a number of compounds. From this information, determine the simplest formula for each compound. Use the density information to determine the correct formulas for these compounds.

1. 27.3% carbon; 72.7% oxygen; 1.0 dm^3 has a mass of 1.96 g

2. 92.3% carbon; 7.7% hydrogen; 500.0 cm^3 have a mass of 1.72 g

3. 94.12% sulfur; 5.88% hydrogen; vapor density = 17.0

4. 47.4% sulfur; 52.6% chlorine; 50.0 mL have a mass of 0.301 g

5. 30.43% nitrogen; 69.57% oxygen; 259 cm^3 have a mass of 1.035 g

6. 95% fluorine; 5% hydrogen; 100.0 mL have a mass of 0.18 g

7. 46.6% nitrogen; 53.4% oxygen; 300.0 cm^3 have a mass of 0.40 g

8. 85.72% carbon; 14.28% hydrogen; 2.0 L have a mass of 2.50 g

9. 37.6% carbon; 12.5% hydrogen; 49.9% oxygen; 60.0 mL have a mass of 0.86 g

10. 10.04% carbon; 0.84% hydrogen; 89.12% chlorine; 1.0 dm^3 has a mass of 5.30 g

11. 46.2% carbon; 53.8% nitrogen; 219 cm^3 have a mass of 0.508 g

12. 42.9% carbon; 57.1% oxygen; 400.0 mL have a mass of 0.50 g

13. 40.00% carbon; 6.67% hydrogen; 53.33% oxygen; 10.0 cm^3 have a mass of 0.027 g

14. 80.0% carbon; 20.0% hydrogen; 150 mL have a mass of 0.20 g

15. 73.8% carbon; 8.7% hydrogen; 17.5% nitrogen; 100.0 cm^3 have a mass of 0.72 g

(continued)

Molecular Formula *(continued)*

Name _____

Date _____

Group II. Given below are the percentage compositions for a number of compounds. From this information, determine the simplest formula for each compound. Use the freezing and boiling point information to determine the correct formulas for these compounds. All of the compounds are non-ionizable organic compounds.

1. benzylamine: 78.5% carbon; 13.1% nitrogen; 8.4% hydrogen; 107 g dissolved in 1000 g of water have a freezing point of –1.86 °C.

2. 2,3-butadione: 55.8% carbon; 7.0% hydrogen; 37.2% oxygen; 86 g dissolved in 1000 g of water have a boiling point of 100.52 °C.

3. t-butyl alcohol: 64.9% carbon; 13.5% hydrogen; 21.7% oxygen; 37 g dissolved in 500 g of water have a freezing point of –1.86 °C.

4. ethyl caprate: 72.0% carbon; 12.0% hydrogen; 16% oxygen; the boiling point of 50 g of this compound in 250 g of water is 100.52 °C.

5. urotropin: 51.4% carbon; 8.6% hydrogen; 40.0% nitrogen; 14 g dissolved in 100 g of water have a boiling point of 100.52 °C.

6. isocrotonic acid: 55.8% carbon; 7.0% hydrogen; 37.2% oxygen; 25.8 g in 300 g of water have a freezing point of –1.86 °C.

7. isosuccinic acid: 40.6% carbon; 5.1% hydrogen; 54.3% oxygen; 29.5 g dissolved in 500 g of water have a freezing point of –0.93 °C.

8. ketine: 67.6% carbon; 7.4% hydrogen; 25.9% nitrogen; 10.8 g dissolved in 400 g of water have a boiling point of 100.13 °C.

9. alpha naphthalenedisulfonic acid: 41.7% carbon; 2.8% hydrogen; 22.2% sulfur; 33.3% oxygen; the freezing point of 43.2 g of this compound dissolved in 600 g of water is –0.465 °C.

10. nicotine: 74.0% carbon; 8.6% hydrogen; 17.3% nitrogen; 12.2 g dissolved in 250 g of water have a freezing point of –0.56 °C.

11. nitralinic acid: 31.3% carbon; 0.9% hydrogen; 55.7% oxygen; 12.2% nitrogen; 11.5 g of this compound dissolved in 150 g of water have a boiling point of 100.17 °C.

(continued)

Chemistry Problems

12. oxalic acid: 26.7% carbon; 2.2% hydrogen; 71.1% oxygen; the freezing point of a solution of 51.3 g of oxalic acid in 200 g of water is –5.30 °C.

13. piperazinium salt of oxalic acid: 41.0% carbon; 6.8% hydrogen; 15.9% nitrogen; 36.4% oxygen; 250 g of this compound dissolved in 300 g of water have a boiling point of 102.45 °C.

14. paraldol: 54.6% carbon; 9.0% hydrogen; 36.4% oxygen; 358 g of this compound dissolved in 400 g of water have a freezing point of –9.50 °C.

15. diacetone alcohol: 62.1% carbon; 10.4% hydrogen; 27.5% oxygen; 56.1 g dissolved in 150 g of water have a freezing point of –6.0 °C.

*In problems involving reduction with **carbon**, **assume the** carbon **is completely oxidized to** carbon dioxide.*

Mass-Mass Problems

1. Carbon dioxide is produced in the reaction between calcium carbonate and hydrochloric acid. How many grams of calcium carbonate would be needed to react completely with 15.0 grams of hydrochloric acid? How many grams of carbon dioxide would be produced in this reaction? How many grams of calcium chloride would be formed at the same time?

2. Sulfur dioxide may be catalytically oxidized to sulfur trioxide. How many grams of sulfur dioxide could be converted by this process if 100.0 g of oxygen are available for the oxidation?

3. Lightning discharges in the atmosphere catalyze the conversion of nitrogen to nitric oxide. How many grams of nitrogen would be required to make 25.0 g of nitric oxide in this way?

4. Iron(III) oxide may be reduced to pure iron either with carbon monoxide or with coke (pure carbon). Suppose that 150.0 kg of ferric oxide are available. How many kilograms of carbon monoxide would be required to reduce the oxide? How many kilograms of coke would be needed? In each case, how many kilograms of pure iron would be produced?

5. Zinc metal will react with either hydrochloric or sulfuric acid to produce hydrogen gas. If 50.0 g of zinc are to be used in the reaction, how much of each acid would be needed to completely react with all of the zinc? How much hydrogen gas (in grams) would be produced in each case?

6. Calcium phosphate, silicon dioxide, and coke may be heated together in an electric furnace to produce phosphorus, as shown in the following equation:

$$2 \ Ca_3(PO_4)_2 + 6 \ SiO_2 + 10C \longrightarrow 6 \ CaSiO_3 + 10 \ CO \uparrow + P_4$$

(continued)

In this reaction, how many kilograms of calcium phosphate would be needed to make 100.0 kg of phosphorus? How much calcium silicate would be formed as a by-product? How much sand would be used up in this same reaction?

7. Phosphoric acid is produced in the reaction between calcium phosphate and sulfuric acid. How much of the phosphoric acid would be produced from 55 g of the calcium phosphate? What other product is formed and in what quantity?

8. Phosphine, PH_3, is formed when calcium phosphide is added to water. How many grams of phosphine can be obtained from 200.0 g of calcium phosphide? How many grams of the other product are formed?

9. Coke will reduce hot arsenious oxide to pure arsenic (As_4). How much coke would be required to completely reduce 500.0 kg of the oxide? How much arsenic is recovered from this reaction?

10. How much iron will be required to release all of the antimony from 10.0 g antimony trisulfide? How much antimony is obtained in this reaction? (Ferrous sulfide is formed.)

11. Sodium tetraborate is produced according to the following reaction:

$$4 \, H_3BO_3 + 2 \, NaOH \rightarrow 7 \, H_2O + Na_2B_4O_7$$

How much boric acid is needed to make 150 kilograms of sodium tetraborate?

12. Calcium oxide can be prepared by heating calcium in oxygen. How much calcium would be needed to make 15.0 g of calcium oxide in this way?

13. If the calcium oxide were prepared by heating calcium carbonate, how much of the carbonate would be required to produce the 15.0 g of the oxide?

14. If the calcium oxide were to be obtained by the heating of calcium hydroxide, how much hydroxide would be needed to obtain the 15.0 g?

15. If the 15.0 g of calcium oxide in the above problems were allowed to react with water, what amount of calcium hydroxide would be produced?

16. How much magnesium sulfate is needed to completely react with 145 g of sodium chloride? How much sodium sulfate could be produced by this reaction?

17. In nature, carbon dioxide in the atmosphere reacts with calcium hydroxide to produce calcium carbonate. This calcium carbonate will react with carbonic acid (carbon dioxide and water) to form calcium bicarbonate. What mass of calcium bicarbonate would eventually be produced by the action of carbon dioxide on 1.0 g of calcium hydroxide?

18. Hard water, containing calcium bicarbonate as hardness, may be softened by adding sodium hydroxide. Sodium carbonate, calcium carbonate, and water all are formed in the reaction. What mass of sodium hydroxide would be needed to react with all the calcium bicarbonate formed in problem 17?

19. Permanent hard water contains calcium sulfate and can be softened by the addition of sodium carbonate. What amount of sodium carbonate would have to be added to react with 1.0 kg of calcium sulfate?

20. Beryllium hydroxide is amphoteric. It reacts with hydrochloric acid to give beryllium chloride and with sodium hydroxide to give sodium beryllate (Na_2BeO_2). If 20.0 grams of beryllium hydroxide are available, how many grams of beryllium chloride can be prepared? How many grams of sodium beryllate could be made? How many grams of hydrochloric acid and sodium hydroxide would be required for these reactions, respectively?

21. The rare earth element terbium may be prepared by the electrolysis of its fused chloride (valence of terbium = 3). How many grams of the chloride would be needed to supply 1.00 g of the pure metal?

(continued)

22. Gadolinium, europium, and samarium may all be prepared in the same way as terbium. How much of the chloride would be required, in each case, to produce 10.0 μg of the pure element? All of these elements have a valence of 3.

23. The metals manganese, chromium, indium, and tin are all prepared by the reduction of their oxides with carbon. How much carbon will be needed to reduce 100.0 kg of each of the following oxides: manganese dioxide, chromium(III) oxide, indium oxide, tin(II) oxide, tin(IV) oxide?

24. Lanthanum and cerium may be obtained in the free state by fusing their respective chlorides with metallic potassium. What amount of the chloride is needed in each case to make 1.0 kg of the metal?

25. How much anhydrous copper(II) sulfate may be obtained by heating 100.0 g of hydrated cupric sulfate?

26. How much mercury metal can be obtained in the reaction between tin(II) chloride and mercury(II) chloride? How much tin(IV) chloride is obtained in the same reaction? Assume that excess tin(II) chloride was permitted to react with 5.0 g of mercuric chloride.

27. Gold will dissolve in the acid mixture known as aqua regia according to the following reaction:

$$Au + HNO_3 + 3\ HCL \rightarrow AuCl_3 + NO \uparrow + 2\ H_2O$$

How much auric chloride will be produced in this reaction when one starts with 5.0 mg of gold? How much nitric acid must be added initially to dissolve all this gold?

28. How many kg of coke (carbon) are needed to reduce 100.0 kg of lead monoxide to free lead?

29. How much silver bromide can be precipitated by the action of 15.0 g of silver nitrate on an excess of sodium bromide?

(continued)

Name _____

Date _____

30. How many grams of sulfuric acid will react exactly with 400.0 g of aluminum metal?

31. 100.0 g of a 50-50 mixture of sodium chloride and potassium nitrate are allowed to react. Which reactant is in excess and by how much?

32. 75 g of potassium hydroxide are permitted to react with 50.0 g of hydrochloric acid. How much potassium chloride is formed?

33. 120 g of sulfuric acid are added to 230.0 g of barium peroxide. Which reactant is in excess and by how much? How much barium sulfate is precipitated in this reaction?

34. 10.0 g of hydrogen and 75 g of oxygen are exploded together in a reaction tube. How much water is produced? What other gas is found in the tube (besides water vapor) after the reaction, and how much of this gas is there?

35. 50.0 g of oxygen are available for the combustion of 25 g of carbon. Is this an adequate amount? If so, by how much in excess is the oxygen? If not, by how much is the carbon in excess?

36. 75.0 g of zinc are added to 120 g of sulfuric acid. How much hydrogen gas is evolved?

37. 100.0 g of lithium metal are dropped into 1.00 dm^3 of water. How much hydrogen is produced?

38. How much ammonia is evolved when 34 g of ammonium chloride are added to 37 g of potassium hydroxide?

39. How many grams of carbon dioxide can be obtained from the action of 100.0 g of sulfuric acid on 100.0 g of calcium carbonate?

(continued)

40. Are 15 g of chlorine gas enough to replace 30.0 g of bromine from a solution of sodium bromide?

41. Will 100.0 g of copper completely replace all the silver in 200.0 g of silver nitrate?

42. Will 45 g of calcium chloride completely precipitate silver chloride from 100.0 g of silver nitrate?

Mass-Volume Problems

1. Hydrogen sulfide may be prepared in the laboratory by the action of hydrochloric acid on iron(II) sulfide. How much iron(II) sulfide would be needed to prepare 15 dm^3 of hydrogen sulfide?

2. When hydrogen sulfide is burned, two different reactions may occur:

$$2\ H_2S + O_2 \rightarrow 2\ H_2O + 2\ S \qquad or \qquad 2\ H_2S + 3\ O_2 \rightarrow 2\ H_2O + 2\ SO_2 \uparrow$$

Assuming that 100.0 g of oxygen were available for this reaction, how much hydrogen sulfide (in cubic decimeters) could be oxidized in each case? How many grams of sulfur would be produced in the first reaction? How many cubic decimeters of sulfur dioxide would be produced in the second reaction?

3. When magnesium burns in air, it reacts with both the oxygen and the nitrogen in the air, forming magnesium oxide and magnesium nitride. If 20.0 dm^3 of air are used in this reaction, what mass of magnesium oxide is formed? What mass of the nitride is produced?

4. How many cubic decimeters of nitrogen gas can be produced by the decomposition of 50.0 g ammonium nitrite? Water is the second product.

5. 250 dm^3 of hydrogen are used to reduce copper(II) oxide. What mass of copper is to be expected?

(continued)

Chemistry Problems

6. The Haber and Claude processes are methods by which ammonia is synthesized from its elements. What mass of ammonia (in kilograms) is to be expected when 4000.0 dm^3 of nitrogen are used in these processes?

7. When ammonia is oxidized in the Ostwald process, nitric acid and water are produced. What mass of nitric acid will be formed when 1500.0 dm^3 of ammonia are used?

8. How many cubic decimeters of oxygen gas can be obtained from the thermal decomposition of 500.0 g of potassium chlorate?

9. What volume of oxygen can be obtained by the addition of excess water to 1.0 g of sodium peroxide?

10. What is the total volume of water gas produced when excess steam is passed over 50.0 g of hot carbon?

11. How many grams of lithium must be added to water in order to obtain 15 dm^3 of hydrogen?

12. How many grams of hydrochloric acid must be added to an excess of zinc arsenide in order to obtain 10.0 dm^3 of arsine (AsH$_3$)?

13. How many grams of arsenic can be obtained from the decomposition of 12 dm^3 of arsine?

14. It is desired to prepare 100.0 g of silicon tetrafluoride by adding hydrogen fluoride to silicon dioxide. How many cubic decimeters of hydrogen fluoride will be required for this process?

15. Pure boron can be obtained by the reduction of boron trichloride with hydrogen. How much hydrogen (in cubic decimeters) is necessary for this process if 500.0 g of boron trichloride are available for the process?

(continued)

Chemistry Problems

16. When black gunpowder explodes, potassium nitrate, carbon, and sulfur react with each other to form nitrogen, carbon dioxide, and potassium sulfide. If the original mixture contains 50.0 g of potassium nitrate, what is the total volume of the gases produced in the reaction?

17. How many cubic decimeters of hydrogen gas are required to convert 5 g of stannic chloride to stannous chloride? The second product is hydrogen chloride.

18. When ammonia is passed over hot calcium, calcium hydride (CaH_2), and free nitrogen gas are formed. If 30.0 dm^3 of nitrogen can be recovered from this reaction, what mass of calcium was originally used?

19. Calcium nitride will react with water to form ammonia and calcium hydroxide. How many grams of the nitride must be used initially in order to prepare 100.0 dm^3 of ammonia?

20. Very hot zinc will react with steam to form zinc oxide and hydrogen. What volume of steam would be necessary in order to use up 20.0 g of zinc completely? What volume of hydrogen would be produced in this case?

Volume-Volume Problems

1. How many cubic decimeters of hydrogen gas are needed to react completely with 50 dm^3 of chlorine gas? How many cubic decimeters of hydrogen chloride are produced in this reaction?

2. What volume of oxygen is needed to completely oxidize 25 dm^3 of carbon monoxide? How much carbon dioxide is produced?

3. 200 dm^3 of ammonia are produced in the reaction between nitrogen and hydrogen. How many cubic decimeters of each of these gases were used up in this reaction?

(*continued*)

4. 5 dm³ of nitrogen and 15 dm³ of oxygen are introduced into a reaction tube and then ignited. What volume of nitric oxide is produced? Which of the two original gases is entirely used up? How much of the other gas remains?

5. Will 30.0 dm³ of fluorine gas completely react with 50.0 L of hydrogen gas? Which gas is in excess and by how much? How much hydrogen fluoride (in dm³) is formed in this reaction?

6. In the preparation of sulfuric acid, sulfur dioxide must be catalytically oxidized to sulfur trioxide. How much oxygen must be available to convert 50.0 dm³ of sulfur dioxide according to this process?

7. Nitric oxide converts spontaneously to nitrogen dioxide by combining with atmospheric oxygen. If nitric oxide escapes from a bottle at the rate of 7 dm³/min, at what rate is nitrogen dioxide being produced?

8. In the preparation of water gas from steam and hot carbon, it is found that carbon monoxide is being formed at the rate of 30 dm³/s. At what rate, then, is the hydrogen being produced in the same reaction?

Combinations

1. Vanadium metal is prepared by reducing vanadium chloride (VCl_2) with hydrogen. How many dm³ of hydrogen are needed to release 1.0 g of vanadium metal? How many grams of vanadium chloride are needed for this operation?

2. The Solvay process is an important commercial method of preparing sodium carbonate. In the first step of this process, ammonia, water, and carbon dioxide react to produce ammonium bicarbonate. This product is then treated in the second step with sodium chloride to allow conversion to sodium bicarbonate. Finally, the sodium bicarbonate is heated, and sodium carbonate, carbon dioxide, and water are produced. If this process is initiated with a stream of ammonia flowing at the rate of 40.0 dm³/s, at what rate must carbon dioxide be supplied in the first step? How much sodium chloride (in grams per second) is required to permit the second step to occur as required? Finally, how much sodium carbonate (in grams) and how much carbon dioxide (in dm³) are formed every second under these conditions?

(continued)

3. 200.0 g of sodium chloride are fused and electrolyzed. How many grams of sodium and how many dm^3 of chlorine are obtained in this process?

4. How many grams of magnesium metal are needed to react completely with 50.0 g of phosphoric acid? How many dm^3 of hydrogen are produced?

5. It is desired to prepare 50.0 g of water by synthesis. How many dm^3 of hydrogen must be used? How many dm^3 of oxygen?

6. An unknown amount of potassium chlorate was heated until no more oxygen was evolved. 15.824 g of potassium chloride remained in the test tube. What mass of potassium chlorate had originally been placed in the tube? What volume of oxygen gas was evolved during the experiment?

7. How many grams of carbon can be completely burned in 15 dm^3 of oxygen? How many dm^3 of carbon dioxide gas are produced in this reaction?

8. What volume of chlorine gas is needed to completely replace 45 g of bromine from sodium bromide? How many grams of sodium chloride are produced simultaneously?

9. Concentrated hydrochloric acid reacts with manganese dioxide to produce chlorine, manganese(II) chloride, and water. How many grams of manganese dioxide are needed in this reaction if it is desired to collect 20.0 dm^3 of chlorine gas? What mass of manganese(II) chloride will be formed?

10. How many grams of magnesium metal are required to liberate 250 cm^3 of hydrogen gas from hydrochloric acid? Exactly how much acid would be used up in this reaction?

(continued)

11. 45 g of mercury(II) oxide are heated until no more oxygen is produced. What volume of oxygen is expected in this case? How much mercury would you expect to collect at the same time?

12. When ammonia reacts with hydrogen chloride (gas), a white solid, ammonium chloride, is formed. How much ammonium chloride would be produced when 500.0 cm^3 of ammonia are used? What volume of hydrogen chloride is needed to carry out this reaction?

 **Ionic Equations**

Name _____

Date _____

Group I. Write five ionic equations to show the process of ionization.

Group II. Write five ionic equations which show the process of dissociation.

Group III. Write either an ionic or a molecular formula, whichever is correct, for each of the following compounds.

1. hydrogen sulfide

2. glycerol

3. lithium chloride

4. potassium hydroxide

5. hydrogen cyanide

6. potassium phosphate

7. beryllium hydroxide

8. calcium iodide

9. hydrogen chloride

10. potassium sulfite

11. ammonium bromide

12. carbon tetrachloride

13. calcium hydroxide

14. strontium bromide

15. sodium fluoride

16. hydrogen iodide

17. sodium dihydrogen phosphate

18. sodium sulphate

19. boric acid

20. sucrose

21. samarium bromide

22. hydrogen selenide

23. potassium bromide

24. methane

25. ammonium chloride

(continued)

Name _____

Date _____

Group IV. Write ionic equations (where they can be used) to show what happens when each of the following substances is put into water.

1. sodium chloride

2. barium hydroxide

3. calcium chloride

4. hydrogen iodide

5. ammonium bromide

6. potassium carbonate

7. sodium hydroxide

8. benzoic acid ($HC_7H_5O_2$)

9. methyl alcohol (CH_3OH)

10. sodium phosphate

11. hydrogen telluride

12. potassium iodide

13. strontium hydroxide

14. ethyl alcohol (C_2H_5OH)

15. hydrogen bromide

16. glycol ($C_2H_4(OH)_2$)

17. iron(II) iodide

18. sulfuric acid

19. magnesium fluoride

20. aluminum chloride

21. disodium hydrogen phosphate

22. acetic acid

23. ammonium sulfate

24. lithium fluoride

25. formic acid ($HCHO_2$)

(continued)

Name _____

Date _____

Group V. Write ionic and net ionic equations for each of the following reactions.

1. potassium chloride + sodium nitrate

2. rubidium hydroxide + hydrochloric acid

3. lithium chlorate + ammonium chloride

4. calcium chloride + sodium carbonate

5. calcium carbonate + cesium chloride

6. ammonium carbonate + hydriodic acid

7. aluminum bromide + cadmium nitrate

8. copper(II) iodide + ammonium sulfate

9. iron(III) sulfate + lead(II) chlorate

10. sodium hydroxide + sulfuric acid

11. strontium chloride + potassium sulfate

12. mercury(I) nitrate + nickel sulfate

(continued)

13. phosphoric acid + ammonium hydroxide

14. potassium hydroxide + ammonium bromide

15. copper(II) sulfate + sodium bromide

16. ammonium bromide + calcium hydroxide

17. calcium nitrate + potassium chloride

18. lithium hydroxide + hydrosulfuric acid

19. iron(III) chloride + sodium sulfide

20. iron(II) acetate + potassium phosphate

21. sodium sulfite + hydrobromic acid

22. ammonium phosphate + magnesium nitrate

23. nickel(II) chloride + sodium carbonate

24. calcium hydroxide + acetic acid

25. zinc chloride + sodium phosphate

(continued)

26. barium carbonate + phosphoric acid

27. nitric acid + strontium hydroxide

28. sulfuric acid + sodium chloride

29. potassium sulfate + ammonium nitrate

30. lead(II) sulfate + sodium bromide

31. zinc chlorate + ammonium sulfide

32. calcium chloride + manganese(II) iodide

33. ammonium sulfate + barium hydroxide

34. potassium iodide + ammonium nitrate

35. mercury(II) chloride + aluminum bromide

Group VI. Use ionic equations to show how hydrolysis will occur when each of the following salts is put into water.

1. ammonium chloride

2. sodium acetate

(continued)

Name ______________________

Date ______________________

3. lithium sulfide

4. ammonium sulfate

5. potassium cyanide

6. rubidium acetate

7. sodium borate

8. ammonium sulfite

9. lithium carbonate

10. ammonium nitrate

11. calcium acetate

12. sodium carbonate

13. ammonium acetate

14. potassium sulfite

15. ammonium bicarbonate

Equilibria

Name _____

Date _____

Group I. Write an equilibrium constant for each of the following reactions.

1. $2 SO_2 + O_2 \rightleftharpoons 2 SO_3$

2. $2 CO + O_2 \rightleftharpoons 2 CO_2$

3. $N_2 + 3 H_2 \rightleftharpoons 2 NH_3$

4. $H_2 + Cl_2 \rightleftharpoons 2 HCl$

5. $2 N_2 + O_2 \rightleftharpoons 2 N_2O$

6. $2 NO + O_2 \rightleftharpoons 2 NO_2$

7. $HC_2H_3O_2 \rightleftharpoons H^+ + C_2H_3O_2^-$

8. $HCN \rightleftharpoons H^+ + CN^-$

(continued)

9. $AgCl \rightleftharpoons Ag^+ + Cl^-$

10. $PbI_2 \rightleftharpoons Pb^{2+} + 2\,I^-$

11. $Bi_2S_3 \rightleftharpoons 2\,Bi^{3+} + 3\,S^{2-}$

12. $Ca_3(PO_4)_2 \rightleftharpoons 3\,Ca^{2+} + 2\,PO_4^{3-}$

Group II. In each of the following, determine the unknown quantity from the information given. The number in parentheses refers to the corresponding reaction in Group I to which you should refer.

1. Find K_{eq} if $[SO_2] = 1.0$; $[O_2] = 1.0$; $[SO_3] = 2.0$ (1.)

2. Find K_{eq} if $[CO] = 0.5$; $[O_2] = 0.5$; $[CO_2] = 2.5$ (2.)

3. Find K_{eq} if $[N_2] = 0.25$; $[H_2] = 0.10$; $[NH_3] = 0.010$ (3.)

4. Find K_{eq} if $[H_2] = 2.0 \times 10^{-3}$; $[Cl_2] = 2.5 \times 10^{-2}$; $[HCl] = 1.5 \times 10^{-3}$ (4.)

5. Find $[O_2]$ if $K_{eq} = 45.0$; $[N_2] = 1.0$; $[N_2O] = 1.0$ (5.)

6. Find $[NO]$ if $[O_2] = 0.10$; $[NO_2] = 0.20$; $K_{eq} = 10.0$ (6.)

7. Find $[N_2]$ if $[H_2] = 1.0 \times 10^{-2}$; $[NH_3] = 2.0 \times 10^{-3}$; $K_{eq} = 1.5 \times 10^{-4}$ (3.)

(continued)

8. Find [CO] if $[O_2] = 1.3 \times 10^{-3}$; $[CO_2] = 2.5 \times 10^{-4}$; $K_{eq} = 3.6 \times 10^{-3}$ (2.)

9. Find K_i if $[HC_2H_3O_2] = 0.10$; $[H^+] = [C_2H_3O_2^-] = 0.0010$ (7.)

10. Find K_i if $[HCN] = 0.0010$; $[H^+] = 0.010$; $[CN^-] = 2.0 \times 10^{-8}$ (8.)

11. Find $[C_2H_3O_2^-]$ if $[HC_2H_3O_2] = 1.5 \times 10^{-2}$; $[H^+] = 2.0 \times 10^{-3}$; $K_i = 1.8 \times 10^{-5}$ (7.)

12. Find $[H^+]$ if $[HCN] = 3.6 \times 10^{-3}$ and $[CN^-] = [H^+]$; $K_i = 5.8 \times 10^{-8}$ (8.)

13. Find K_{sp} if the solubility of silver chloride is 4.3×10^{-6} g/100 mL   (9.)

14. Find K_{sp} if the solubility of bismuth sulfide is 2.9×10^{-5} g/100 mL   (11.)

15. Find $[Pb^{2+}]$ if K_{sp} for PbI_2 is 7.5×10^{-9} (10.)

16. Find the solubility in grams/100 mL of calcium ion if K_{sp} for calcium phosphate is 3.2×10^{-24} (12.)

Group III. Solve each of the following problems involving equilibria.

1. Calculate the equilibrium constant for the following reaction, $2A + B \rightleftharpoons 3C + D$, where molar concentrations are A = 3; B = 2; C = 2; and D = 4.

2. The equilibrium constant, K, for the reaction $A + B \rightleftharpoons 2C$ is 50. After mixing equimolar quantities of A and B, the equilibrium concentration of C is found to be 0.50. What are the concentrations of A and B at equilibrium?

(continued)

3. Consider the reaction PCl_5 (g) $\rightleftharpoons$ PCl_3 (g) + Cl_2 (g). The equilibrium mixture in a 9.0 dm^3 container was found to include 0.25 mol PCl_5, 0.36 mol PCl_3, and 0.36 mol Cl_2. From this data, calculate the equilibrium constant for the dissociation at the reaction temperature of 225 °C.

4. Nitrogen is caused to react with hydrogen to form ammonia at 450 °C in a 4.0 dm^3 vessel. At equilibrium, the partial pressures observed for each of the species in the reaction was as follows: NH_3: 900 mm; N_2: 180 mm; and H_2: 305 mm. From this information, calculate the equilibrium constant for the reaction at this temperature.

5. Consider the reaction $H_2 + I_2 \rightleftharpoons 2HI$. The equilibrium constant for this reaction is 32. If, at equilibrium, the concentration of HI is 0.40 and that of I_2 is 0.05, what is the concentration of H_2?

6. The equilibrium constant for the following reaction is 8.

$$4 Al (s) + 3O_2 (g) \rightleftharpoons 2Al_2O_3 (s)$$

What is the concentration of oxygen at equilibrium?

7. For the reaction $H_2 + I_2 \rightleftharpoons 2HI$, what is the equilibrium constant if the following concentration of substances is observed at equilibrium?

$[H_2] = 5.62$ M $\qquad\qquad$ $[HI] = 7.89$ M

$[I_2] = 0.130$ M

8. Given the equilibrium concentrations shown below, what is the dissociation constant for ammonia?

$[NH_3] = 0.0015$ M $\qquad\qquad$ $[N_2] = 0.069$ M

$[H_2] = 0.032$ M

9. In a saturated solution of $Ag_2C_2O_4$, the concentration of silver ion is 3.4×10^{-4} mol/dm^3. From this information, compute the solubility product of $Ag_2C_2O_4$.

(continued)

10. What amount of chloride ion must be exceeded before silver chloride will precipitate out of a solution in which the concentration of silver ion is 3.6×10^{-3} mol/dm^3? K_{sp} for AgCl is 1.8×10^{-10} at 25 °C.

11. A solution in contact with solid silver chromate contains CrO_4^{-2} ions in the concentration of 0.099 mol/L and Ag^{+1} ions at a concentration of 4.0×10^{-6} mol/dm^3. From this information, what is the value of K_{sp} for silver chromate?

12. K_{sp} for HgS is 3×10^{-53}. If mercury(II) nitrate has a concentration of 1.0×10^{-14} mol/dm^3, what is the maximum amount of sulfide ion that can exist in this solution?

13. Consider the reaction $SrCO_3$ (s) $\rightleftharpoons$ Sr^{2+} + CO_3^{2-}. Given that the molar concentrations of the two ions in this reaction are each 3.07×10^{-5} mol/dm^3, what is the value of K_{sp} for $SrCO_3$?

14. If the solubility product for AgBr is 4.9×10^{-13} at 25 °C, and the molar concentrations of Ag^+ and Br^- are each 9.1×10^{-5} mol/dm^3 in a solution of AgBr, will there be a precipitate of AgBr formed?

15. Given that K_{sp} for $Ca_3(PO_4)_2$ is 1.0×10^{-26}, what are the molar concentrations of Ca^{2+} and PO_4^{3-} in a saturated solution?

16. The solubility product of $PbCl_2$ at 25 °C is 1.6×10^{-5}. If $[Cl^-] = 3.0 \times 10^{-2}$, what is the concentration of lead ion in equilibrium with the chloride ion?

17. **Le Chatelier's Principle**
    Predict the effects of (a) an increase in pressure and (b) an increase in temperature on each of the following reactions.
    1. $2 SO_2$ (g) + O_2 (g) $\rightleftharpoons$ $2 SO_3$ (g)
    2. $2 O_3$ (g) $\rightleftharpoons$ $3 O_2$ (g)
    3. $2 NO_2$ (g) $\rightleftharpoons$ N_2O_4 (g)
    4. CO_2 (g) + H_2 (g) $\rightleftharpoons$ CO (g) + H_2O (g)
    5. C (s) + H_2O (g) $\rightleftharpoons$ CO (g) + H_2 (g)
    6. N_2 (g) + O_2 (g) $\rightleftharpoons$ 2 NO (g)

Group I. Determine the hydrogen ion concentration ($[H^+]$), hydroxide ion concentration ($[OH^-]$), pH and pOH of each of the following solutions. For weak acids and bases, the degree of ionization is given in parentheses.

1. 0.10 M HCl

2. 1.0 M HNO_3

3. 0.10 M NaOH

4. 1.0 M KOH

5. 0.10 M $HC_2H_3O_2$ (1.0%)

6. 0.010 M $HC_2H_3O_2$ (10.0%)

7. 1.0 M $HC_2H_3O_2$ (0.10%)

8. 0.10 M NH_4OH (1.0%)

9. 0.0010 M NH_4OH (10.0%)

10. 0.010 M NH_4OH (1.0%)

11. 0.10 M HCN (0.50%)

12. 0.010 M HCN (1.5%)

13. 1.0×10^{-4} M HCN (3.5%)

14. 0.015 M HCN (1.2%)

15. 0.0038 M HCN (2.3%)

16. 0.0016 M tartaric acid (0.042%)

(continued)

17. 2.1×10^{-4} M tartaric acid $(1.5 \times 10^{-2}\%)$

18. 1.6×10^{-3} M oxalic acid (0.166%)

19. 5.8×10^{-2} M oxalic acid (0.086%)

20. 4.25×10^{-4} M oxalic acid (0.835%)

21. 0.64 M NH_4OH (1.2%)

22. 0.098 M NH_4OH (1.6%)

23. 3.4×10^{-3} M NH_4OH (3.7%)

24. 0.88 M NH_4OH (0.95%)

25. What is the hydrogen ion concentration in a solution containing 0.020 M benzoic acid and 0.010 M sodium benzoate? K_a for benzoic acid is 6.6×10^{-5}, and the acid ionizes as follows:

$$C_6H_5COOH(aq) \rightarrow H^+(aq) + C_6H_5COO^-(aq)$$

26. Calculate the pH of a solution whose hydrogen ion concentration is 0.0036 mol/dm^3.

27. What would be the ionization constant for acetic acid at 25 °C if a 0.010 M solution of this acid is ionized to the extent of 1.38%?

28. 30 g of acetic acid is dissolved in enough water to make one liter of solution. What is the molarity of the solution, and what is the hydrogen ion concentration in the solution?

29. What are the $[H^+]$ and $[ClO^-]$ for a 0.048 M solution of HClO? K_i for HClO is 3.5×10^{-8}.

30. What is the pH of a solution of 0.001 0 M hydrochloric acid?

31. Calculate the pH of a 0.30 M solution of aniline, $C_6H_5NH_2$. K_b for aniline is 4.2×10^{-10}.

(continued)

Chemistry Problems

32. Complete the equations for each of the following acid-base reactions.

   (a) $HF(aq) + CH_3COO^-(aq) \rightarrow$

   (b) $H_2S(aq) + CO_3^{-2}(aq) \rightarrow$

   (c) $NH_4^+(aq) + F^-(aq) \rightarrow$

33. Define acid equilibrium expressions for each of the following reactions.

   (a) $HNO_2(aq) + NH_3(aq) \rightarrow NO_2^-(aq) + NH_4^+(aq)$

   (b) $C_6H_5COOH(aq) + CH_3COO^-(aq) \rightarrow C_6H_5COO^-(aq) + CH_3COOH(aq)$

34. Identify the conjugate acid-base pairs in each of the following reactions.

   (a) $HSO_4^-(aq) + C_2O_4^{2-}(aq) \rightleftharpoons SO_4^{2-} + HC_2O_4^-(aq)$

   (b) $H_2PO_4^-(aq) + HCO_3^-(aq) \rightleftharpoons HPO_4^{2-}(ag) + H_2CO_3(aq)$

   (c) $H_2O + S^{2-}(aq) \rightleftharpoons HS^-(aq) + OH^-(aq)$

   (d) $CN^-(aq) + HC_2H_3O_2(aq) \rightleftharpoons HCN(aq) + C_2H_3O_2^-(aq)$

   (e) $HNO_2(aq) + H_2O \rightleftharpoons H_3O^+(aq) + NO_2^-(aq)$

35. Classify each of the following as a Lewis acid or base.

   (a) H_2O

   (b) O^{2-}

   (c) S^{2-}

   (d) H^+

   (e) BF_3

   (f) Cu^{2+}

   (g) Cl

   (h) SO_3 in the reaction:
   $$SO_3 + CaO \rightleftharpoons CaSO_4$$

Group II. Resolve the following acid base titrations.

1. A volume of 10 mL of 0.75 M NaOH neutralizes a 30 mL sample of HClO solution. What is the concentration of HClO?

2. A volume of 30 mL of 0.25 M HCl neutralizes a 50 mL sample of KOH solution. What is the concentration of KOH?

(continued)

3. A volume of 37 mL of 0.36 M KCN neutralizes a 75 mL sample of HClO solution. What is the concentration of HClO?

4. A volume of 60 mL of 0.60 M HBr neutralizes an 80 mL sample of $Ca(OH)_2$ solution. What is the concentration of $Ca(OH)_2$?

5. A volume of 135 mL of 0.40 M HCl neutralizes a 90 mL sample of $Ca(OH)_2$ solution. What is the concentration of $Ca(OH)_2$?

6. A volume of 90 mL of 0.2 M HBr neutralizes a 60 mL sample of NaOH solution. What is the concentration of the NaOH solution?

7. A volume of 46 mL of 0.40 M NaOH neutralizes an 80 mL sample of HCN solution. What is the concentration of HCN?

8. A volume of 20 mL of 0.25 M $Al(OH)_3$ neutralizes a 75 mL sample of H_2SO_4 solution. What is the concentration of H_2SO_4?

9. A volume of 50 mL of 0.30 M HCl neutralizes a 60 mL sample of $Ca(OH)_2$ solution. What is the concentration of $Ca(OH)_2$?

10. A volume of 9.0 mL of 0.70 M NH_3 neutralizes a 35 mL sample of $HClO_4$ solution. What is the concentration of $HClO_4$?

Standard Solutions

Name _____

Date _____

Standard Solutions

Group I. A. Tell how you would prepare each of the following solutions.

1. 1.0 dm^3 of 1.0 M potassium chloride

2. 1.0 L of 1.0 M strontium nitrate

3. 1.0 dm^3 of 1.0 M sodium sulfate

4. 2.0 L of 1.0 M lithium bromide

5. 5.0 dm^3 of 1.0 M aluminum nitrate

6. 1.0 L of 0.10 M calcium chloride

7. 1.0 dm^3 of 0.10 M beryllium nitrate

8. 1.0 L of 0.20 M sodium nitrate

9. 1.0 dm^3 of 0.010 M magnesium hydroxide

10. 500. mL of 0.10 M ammonium carbonate

11. 40.0 cm^3 of 2.0 M rubidium fluoride

12. 600. mL of 0.10 M sodium sulfide

13. 100. cm^3 of 0.20 M aluminum sulfate

14. 250. mL of 0.040 M ammonium chloride

15. 50. cm^3 of 5.0 M potassium phosphate

16. 400. mL of 0.10 M barium nitrate

17. 500. cm^3 of 6.0 M sodium phosphate

(continued)

Chemistry Problems

18. 25 mL of 0.40 M cesium nitrate

19. 100. cm^3 of 0.50 M mercury(II) chlorate

20. 750. mL of 6.0 M potassium sulfate

21. 1.0 L of 1.0 N hydrochloric acid

22. 1.0 dm^3 of 1.0 N calcium iodide

23. 1.0 L of 1.0 N aluminum chloride

24. 1.0 dm^3 of 1.0 N phosphoric acid

25. 1.0 L of 2.0 N barium acetate

26. 1.0 L of 2.0 N sodium hydroxide

27. 1.0 L of 2.0 N nickel(II) chloride

28. 1.0 L of 4.0 N aluminum nitrate

29. 1.0 L of 6.0 N nitric acid

30. 1.0 dm^3 of 0.10 N zinc chloride

31. 1.0 dm^3 of 0.10 N barium hydroxide

32. 1.0 dm^3 of 0.10 N cadmium sulfate

33. 1.0 dm^3 of 0.10 N hydrobromic acid

34. 1.0 dm^3 of 0.20 N copper(II) bromide

35. 1.0 L of 0.20 N magnesium nitrate

36. 1.0 L of 0.010 N rubidium hydroxide

37. 1.0 L of 0.010 N nickel nitrate

(continued)

Standard Solutions *(continued)*

Name _____

Date _____

38. 1.0 dm^3 of 0.010 N strontium hydroxide

39. 1.0 dm^3 of 0.020 N bismuth bromide

40. 1.0 dm^3 of 0.040 N silver nitrate

41. 2.0 L of 1.0 N aluminum sulfate

42. 4.0 L of 1.0 N copper(II) nitrate

43. 500. cm^3 of 1.0 N hydrogen iodide

44. 200. cm^3 of 1.0 N iron(II) chloride

45. 750 cm^3 of 1.0 N potassium hydroxide

46. 100. cm^3 of 1.0 N mercury(II) chloride

47. 250. cm^3 of 2.0 N cadmium chlorate

48. 300. mL of 0.10 N sulfuric acid

49. 500. mL of 0.10 N calcium chloride

50. 100. mL of 0.10 N nickel sulfate

51. 250. mL of 0.20 N ammonium hydroxide

52. 50.0 mL of 4.0 N iron(III) nitrate

53. 25 cm^3 of 0.80 N aluminum iodide

54. 30.0 cm^3 of 0.010 N lead(II) acetate

55. 200. cm^3 of 0.50 N lithium hydroxide

56. 350. cm^3 of 0.030 N magnesium chloride

57. 150. mL of 0.020 N copper(II) sulfate

(continued)

Chemistry Problems

58. 50.0 mL of 2.0 N barium nitrate

59. 250. mL of 0.20 N manganese(II) sulfate

60. 8.0 L of 0.10 N cadmium nitrate

Group I. B. What is the molarity and normality of each of the following solutions?

1. 40. g of sodium hydroxide in 1.0 dm^3 of solution

2. 122.5 g of potassium chlorate in 1.0 dm^3 of solution

3. 133.5 g of aluminum chloride in 1.0 dm^3 of solution

4. 142 g of sodium sulfate in 1.0 L of solution

5. 342 g of aluminum sulfate in 1.0 L of solution

6. 58.5 g sodium chloride in 1.0 L of solution

7. 98 g of sulfuric acid in 0.50 dm^3 of solution

8. 21.2 g of lithium chloride in 4.0 L of solution

9. 7.8 g of potassium hydroxide in 500. cm^3 of solution

10. 32.7 g of phosphoric acid in 250. cm^3 of solution

11. 49 g of ammonium bromide in 2.0 L of solution

12. 1.2 g of lithium hydroxide in 100. cm^3 of solution

13. 9.0 g of calcium chloride in 250. mL of solution

14. 20.0 g of nitric acid in 500. mL of solution

15. 12.3 g of aluminum sulfate in 400. mL of solution

16. 1.2 g of hydrochloric acid in 750. cm^3 of solution

(continued)

17. 4.1 g of manganese(II) chloride in 300. cm^3 of solution

18. 17.2 g of potassium phosphate in 250. cm^3 of solution

19. 6.6 g of barium hydroxide in 50. mL of solution

20. 44.2 g of ammonium sulfate in 600. mL of solution

21. 0.014 g of calcium nitrate in 10.0 mL of solution

22. 0.28 g of hydrobromic acid in 45 cm^3 of solution

23. 4.3 g of sodium nitrite in 100. cm^3 of solution

24. 2.9 g of ferric sulfate in 250. cm^3 of solution

25. 0.40 g of strontium hydroxide in 38 mL of solution

Percent Solutions

Group II. Express the concentration of each of the following solutions in percent mass/ volume (m/v), mass/mass (m/m), or volume/volume (v/v) as indicated.

1. 5.0 g in 100.0 cm^3 (m/v)

2. 2.5 g in 100.0 g (m/m)

3. 1.0 mL in 100.0 mL (v/v)

4. 3.0 g in 50.0 cm^3 (m/v)

5. 0.25 g in 25.0 mL (m/v)

6. 1.5 g in 50.0 g (m/m)

7. 0.75 mL in 10.0 cm^3 (v/v)

8. 1.89 g in 30.0 mL (m/v)

9. 0.62 g in 45.0 (m/m)

10. 15.3 mL in 2.65 L (v/v)

(continued)

11. 4.5 g in 138.6 mL (m/v)

12. 0.575 g in 265 g (m/m)

Group III. What mass of solute is contained in each of the following solutions?

1. 100.0 mL of a 1.5% (m/v) solution

2. 100.0 g of a 2.3% (m/m) solution

3. 50.0 cm^3 of a 2.5% (m/v) solution

4. 25.0 mL of a 0.50% (m/v) solution

5. 75.0 g of a 1.2% (m/m) solution

6. 50.0 mL of a 2.8% (m/v) solution

7. 32.5 cm^3 of a 1.8% (m/v) solution

8. 75.0 mL of a 2.6% (m/v) solution

9. 56.5 g of a 1.4% (m/m) solution

10. 11.2 mL of a 0.45% (m/v) solution

11. 1.45 mL of a 0.33% (m/v) solution

12. 2.43 L of a 0.042% (m/v) solution

Group IV. What volume of each of the following solutions would be needed to deliver the stated amount of solute?

1. 1.0 g of a 1.0% (m/v) solution

2. 3.0 g of a 1.5% (m/v) solution

3. 0.5 g of a 1.0% (m/v) solution

4. 1.5 g of a 0.50% (m/v) solution

5. 0.25 g of a 2.0% (m/v) solution

6. 1.0 g of a 0.9% (m/v) solution

7. 2.5 g of a 0.45% (m/v) solution

(continued)

8. 2.0 cm^3 of a 1.5% (v/v) solution

9. 10.0 mL of a 2.5% (v/v) solution

10. 3.5 mL of a 1.5% (v/v) solution

11. 5.0 g of a 0.9% (m/v) solution

12. 4.2 cm^3 of a 1.3% (v/v) solution

Group V. Calculate the normality of the following solutions from the information given (SG = specific gravity of the solution; % = percentage concentration of the solution).

1. hydrochloric acid: SG = 1.19;   % = 36

2. nitric acid: SG = 1.42;   % = 69.5

3. sulfuric acid: SG = 1.84;   % = 96.0

4. ammonium hydroxide: SG = 0.90;   % = 58.6

5. acetic acid: SG = 1.06;   % = 99.5

6. cupric sulfate: SG = 1.20;   % = 18

7. magnesium chloride: SG = 1.27;   % = 30.0

8. nickel(II) nitrate: SG = 1.38;   % = 35

9. sodium thiosulfate: SG = 1.30;   % = 31.9

10. zinc chloride: SG = 1.96;   % = 70.0

Group VI. If you were given 100. mL of each of the following solutions, how would you make a 0.10 N solution of hydrochloric acid?

1. 1.0 N HCl

2. 6.0 N HCl

3. 5.0 N HCl

4. 2.0 N HCl

5. 2.4 N HCl

6. 3.0 N HCl

7. 10.0 N HCl

8. 0.50 N HCl

9. 0.40 N HCl

10. 1.0 M HCl

(continued)

Group VII. If you were given 100. cm³ of each of the following solutions, how would you make a 0.0020 N solution of sulfuric acid?

1. 0.10 N H_2SO_4

2. 4.0 N H_2SO_4

3. 36 N H_2SO_4

4. 6.0 N H_2SO_4

5. 0.010 N H_2SO_4

6. 0.50 N H_2SO_4

7. 0.020 N H_2SO_4

8. 2.0 M H_2SO_4

9. 0.20 N H_2SO_4

10. 0.40 M H_2SO_4

Titrations

Group VIII. Solve the following problems.

1. You are given 40.0 mL of 0.10 N hydrochloric acid. What volume of sodium hydroxide is needed to neutralize completely all the acid when the base has the following normalities:

(a) 0.20 N

(b) 0.010 N

(c) 4.0 N

(d) 5.0 N

(e) 0.002 0 N

(f) 1.0 N

(g) 0.40 N

(h) 2.0 N

(i) 0.10 N

(j) 0.50 N

2. You are given 50.0 cm³ of 0.20 N sodium hydroxide. If the following volumes of acid were used to neutralize this base, what was the normality of the acid in each case?

(a) 10.0 mL

(b) 250. cm³

(c) 5.0 mL

(d) 100.0 cm³

(e) 40.0 mL

(f) 70.0 mL

(g) 20.0 mL

(h) 62.5 cm³

(i) 12.5 cm³

(j) 30.0 cm³

(continued)

3. What volume of 0.50 N sulfuric acid is needed to react completely with 10.0 cm^3 of 2.0 N potassium hydroxide?

4. 25 mL of acetic acid exactly neutralize 30.0 mL of 0.40 N sodium hydroxide. What is the normality of the acetic acid?

5. 23.8 mL of sulfuric acid neutralize 34.2 mL of 0.20 N potassium hydroxide. What is the normality of the sulfuric acid?

6. How many milliliters of 2.0 N sodium hydroxide are needed to neutralize 15 mL of 0.60 N hydrochloric acid?

7. What volume of 0.50 N ammonium hydroxide is required to react completely with 15 cm^3 of 0.2 N hydrochloric acid?

8. 16 mL of 2.0 N sodium hydroxide neutralize 25 mL of hydrochloric acid. What is the normality of the acid?

9. 50.0 mL of 0.010 N acetic acid will neutralize 65.0 mL of potassium hydroxide. What is the normality of the base?

10. How many milliliters of 0.10 N sodium hydroxide are needed to react completely with 26.2 cm^3 of 0.30 N sulfuric acid?

(continued)

Name _____

Date _____

11. How many milliliters of 0.10 N hydrochloric acid are needed to react completely with the following?

  (a) 5.0 g of sodium hydroxide

  (b) 100.0 mg of calcium hydroxide

  (c) 12.5 g of potassium hydroxide

  (d) 2.00 g of lithium hydroxide

  (e) 10.0 mg of barium hydroxide

  (f) 30.0 g of sodium hydroxide

  (g) 6.2 g of potassium hydroxide

  (h) 15.7 g of barium hydroxide

  (i) 1.0 g of calcium hydroxide

  (j) 40.0 mg of strontium hydroxide

12. 50.0 mL of a hydrochloric acid solution are required to react completely with 1.204 g of calcium carbonate. What is the normality of the acid?

13. 42.6 cm^3 of sulfuric acid precipitate exactly 3.260 g of barium sulfate from a solution. What is the normality of the acid?

14. A 0.15 N solution of calcium chloride is added to a solution of ammonium carbonate and 2.010 g of calcium carbonate are precipitated. What volume of chloride solution was added?

15. 3.0 g of silver nitrate are dissolved in water. 36.0 mL of a sodium chloride solution are needed to completely precipitate all the silver chloride from this solution. What was the normality of the sodium chloride solution?

(*continued*)

16. An excess of hydrogen sulfide gas is passed into 40.0 mL of each of the following solutions: bismuth nitrate, cadmium nitrate, cupric nitrate, ferrous nitrate, ferric nitrate, lead nitrate, manganese nitrate, mercurous nitrate, mercuric nitrate, nickel nitrate, silver nitrate, zinc nitrate. The following amounts of precipitates are formed. What was the normality of the original nitrate solution in each case?

(a)	Bi_2S_3	3.24 g	(g) MnS	0.904 g
(b)	CdS	0.834 g	(h) Hg_2S	3.04 g
(c)	CuS	0.654 g	(i) HgS	2.15 g
(d)	FeS	0.215 g	(j) NiS	0.832 g
(e)	Fe_2S_3	1.248 g	(k) Ag_2S	1.462 g
(f)	PbS	1.044 g	(l) ZnS	0.058 g

Molal Solutions

Group IX. Solve the following problems.

1. How would you make each of the following solutions?

    (a) 1.0 m NaCl

    (b) 2.0 m H_2SO_4

    (c) 0.25 m sucrose ($C_{12}H_{22}O_{11}$)

    (d) 500. mL of 1.0 m ethyl alcohol (C_2H_5OH)

    (e) 500. mL of 0.50 m H_3PO_4

    (f) 1.5 dm^3 of 0.25 m nitric acid

    (g) 75 mL of 0.30 m sodium hydroxide

    (h) 150. cm^3 of 0.500 m ammonium hydroxide

2. What is the molality of each of the following solutions?

    (a) 40.0 g of NaOH in 1.00 kg of water

    (b) 45.0 g of glucose ($C_6H_{12}O_6$) in 1.00 kg of water

    (c) 10.0 g of NaCl in 500. g of water

    (d) 30.0 g of sucrose in 125 g of water

    (e) 2.5 g of sulfuric acid in 200. g of water

    (f) 4.5 g of glucose in 200. g of water

    (g) 30.0 g of phosphoric acid in 250. g of water

    (h) 41 g of calcium chloride in 300. mL of water

Colligative Properties

Name _____

Date _____

Solve the following problems.

1. What is the boiling point of a solution that consists of 1.0 mol of sucrose ($C_{12}H_{22}O_{11}$) dissolved in 1.0 kg of water?

2. What is the freezing point of a solution that contains 0.50 mol of sucrose dissolved in 1.0 kg of water?

3. What is the boiling point of a solution that consists of 15.0 g of sucrose in 150. g of water?

4. What is the freezing point of a solution that contains 10.0 g of sucrose in 85.0 g of water?

5. What is the freezing point of a solution that contains 4.50 g of p-cresyl acetate ($C_9H_{10}O_2$) dissolved in 50.0 g of benzene?

6. 35.0 g of an organic compound dissolved in 500. g of water have a freezing point of –3.72 °C. What is the molecular weight of the solute?

7. 63.2 g of an organic compound dissolved in 350. g of water have a boiling point of 101.5 °C. What is the molecular weight of the solute?

8. 5.85 g of an organic compound dissolved in 75.0 g of water have a freezing point of –1.04 °C. What is the molecular weight of the compound?

9. 3.33 g of an organic compound dissolved in 150. g of carbon tetrachloride have a boiling point of 78.5 °C. What is the molecular weight of this compound?

10. An organic compound whose molecular weight is 140 is dissolved in 1000 g of cyclohexane. The freezing point of the solution is 4.75 °C. How much of the compound was dissolved in the cyclohexane?

Balance the following redox reactions.

1. $H_2SO_3 + I_2 + H_2O \rightarrow H_2SO_4 + HI$

2. $FeCl_3 + SnCl_2 \rightarrow FeCl_2 + SnCl_4$

3. $SbCl_5 + KI \rightarrow SbCl_3 + KCl + I_2$

4. $CdS + I_2 + HCl \rightarrow CdCl_2 + HI + S$

5. $KClO_3 + FeSO_4 + H_2SO_4 \rightarrow KCl + Fe_2(SO_4)_3 + H_2O$

6. $CuSO_4 + KI \rightarrow CuI + K_2SO_4 + I_2$

7. $AuCl_3 + KI \rightarrow AuCl + KCl + I_2$

8. $SnCl_2 + HgCl_2 \rightarrow SnCl_4 + HgCl$

9. $Ti_2(SO_4)_3 + Fei_2(SO_4)_3 \rightarrow Ti(SO_4)_2 + FeSO_4$

10. $NaClO + H_2S \rightarrow NaCl + H_2SO_4$

11. $Fe_3O_4 + KMnO_4 + H_2SO_4 \rightarrow Fe_2(SO_4)_3 + K_2SO_4 + MnSO_4 + H_2O$

12. $NaCrO_2 + NaOH + H_2O_2 \rightarrow Na_2CrO_4 + H_2O$

13. $Ag + HNO_3 \rightarrow AgNo_3 + NO + H_2O$

14. $Ag_2S + HNO_3 \rightarrow AgNO_3 + NO + S + H_2O$

(continued)

Chemistry Problems

15. $FeS + HNO_3 \rightarrow Fe(NO_3)_3 + NO + S + H_2O$

16. $AgNO_3 + NaClO \rightarrow AgCl + AgClO_3 + NaNO_3$

17. $Ca(OH)_2 + Cl_2 \rightarrow CaCl_2 + Ca(ClO_3)_2 + H_2O$

18. $MnO_2 + FeSO_4 + H_2SO_4 \rightarrow MnSO_4 + Fe_2(SO_4)_3 + H_2O$

19. $I_2 + Na_2S_2O_3 \rightarrow Na_2S_4O_6 + NaI$

20. $Bi(NO_3)_3 + Al + NaOH \rightarrow Bi + NH_3 + NaAlO_2$

21. $Cu_2As_2O_7 + Zn + H_2SO_4 \rightarrow Cu + AsH_3 + ZnSO_4 + H_2O$

22. $FeCl_2 + HCl + HNO_3 \rightarrow FeCl_3 + NO + H_2O$

23. $PbCrO_4 + HCl \rightarrow PbCl_2 + CrCl_3 + Cl_2 + H_2O$

24. $K_2Cr_2O_7 + HCl + H_2S \rightarrow KCl + CrCl_3 + S + H_2O$

25. $Ag + HNO_3 + Ca(ClO)_2 \rightarrow AgCl + Ca(NO_3)_2 + H_2O$

26. $HgS + HCl + HNO_3 \rightarrow HgCl_2 + S + NO + H_2O$

27. $Na_2Cr_2O_7 + HNO_3 + H_2O_2 \rightarrow H_3CrO_8 + NaNO_3 + H_2O$

28. $K_2MnO_4 + HNO_3 \rightarrow KMnO_4 + MnO_2 + KNO_3 + H_2O$

29. $KMnO_4 + NaNO_2 + HCl \rightarrow KCl + MnCl_2 + NaNO_3 + H_2O$

30. $MnCl_2 + NaOH + Br_2 \rightarrow MnO_2 + NaCl + NaBr + H_2O$

(continued)

31. $MnO_2 + HNO_3 + H_2O_2 \rightarrow Mn(NO_3)_2 + O_2 + H_2O$

32. $Mn(NO_3)_2 + KIO_4 + H_2O \rightarrow HMnO_4 + KIO_3 + HNO_3$

33. $Mn(NO_3)_2 + Pb_3O_4 + HNO_3 \rightarrow HMnO_4 + Pb(NO_3)_2 + H_2O$

34. $HMnO_4 + HNO_3 \rightarrow Mn(NO_3)_2 + H_2O + O_2$

35. $Cu(AsO_2)_2 + HCl + HNO_3 + H_2O \rightarrow CuCl_2 + H_2AsO_4 + NO$

36. $H_2C_2O_4 + Ce(SO_4)_2 \rightarrow CO_2 + H_2SO_4 + Ce_2(SO_4)_3$

37. $Cl_2 + AgNO_3 + H_2O \rightarrow AgCl + HClO_3 + HNO_3$

38. $CoSO_4 + KI + KIO_3 + H_2O \rightarrow Co(OH)_2 + K_2SO_4 + I_2$

39. $Fe_3O_4 + HNO_3 \rightarrow Fe(NO_3)_3 + NO + H_2O$

40. $Hg_2I_2 + HCl + HNO_3 \rightarrow HgCl_2 + HIO_3 + NO + H_2O$

41. $ZnI_2 + K_2S_2O_8 \rightarrow I_2 + ZnSO_4 + K_2SO_4$

42. $Mn(NO_3)_2 + HNO_3 + NaBiO_3 \rightarrow HMnO_4 + NaNO_3 + Bi(NO_3)_3 + H_2O$

43. $FeS_2 + Na_2O_2 \rightarrow Fe_2O_3 + Na_2SO_4 + Na_2O$

44. $H_2O_2 + Ce(SO_4)_2 \rightarrow Ce_2 (SO_4)_3 + H_2SO_4 + O_2$

45. $CoCl_2 + HgO + KMnO_4 + H_2O \rightarrow Co(OH)_3 + HgCl_2 + MnO_2 + KCl + H_2O$

46. $UO_2(NO_3)_2 + KI + KIO_3 + H_2O \rightarrow UO_2(OH)_2 + KNO_3 + I_2$

(continued)

Chemistry Problems

47. $V_2O_5 + KI + HCl \rightarrow V_2O_4 + KCl + I_2 + H_2O$

48. $Ca(IO_3)_2 + KI + HCl \rightarrow CaCl_2 + KCl + I_2 + H_2O$

49. $PbO_2 + H_2O_2 + HNO_3 \rightarrow Pb(NO_3)_2 + H_2O + O_2$

50. $CaCl_2O + KI + HCl \rightarrow CaCl_2 + KCl + I_2 + H_2O$

51. $Sn + HNO_3 + H_2O \rightarrow H_2SnO_3 + NO$

52. $Mn(NO_3)_2 + Ca(ClO)_2 + NaOH \rightarrow NaMnO_4 + NaNO_3 + CaCl_2 + H_2O$

53. $KBr + Fe_2(SO_4)_3 \rightarrow Br_2 + K_2SO_4 + FeSO_4$

54. $Sb_2S_5 + HCl + HNO_3 \rightarrow SbCl_5 + S + NO + H_2O$

55. $KMnO_4 + HCl \rightarrow KCl + MnCl_2 + Cl_2 + H_2O$

56. $As + HNO_3 + H_2O \rightarrow H_3AsO_4 + NO$

57. $As_2S_5 + HNO_3 + H_2O \rightarrow H_3AsO_4 + S + NO$

58. $Sb_2S_5 + HCl \rightarrow SbCl_3 + S + H_2S$

59. $CaCl_2O + NaAsO_2 + NaOH \rightarrow CaCl_2 + Na_3AsO_4 + H_2O$

60. $MoO_3 + Zn + H_2SO_4 \rightarrow Mo_2O_3 + ZnSO_4 + H_2O$

61. $KIO_3 + H_2 \rightarrow KI + H_2O$

62. $Na_2Cr_2O_7 + S \rightarrow Na_2SO_4 + Cr_2O_3$

(continued)

63. $ZnSO_4 + KI + KIO_3 + H_2O \rightarrow Zn_5(OH)_8SO_4 + K_2SO_4 + I_2$

64. $MnO_2 + KOH + O_2 \rightarrow K_2MnO_4 + H_2O$

65. $MoO_3 + KI + HCl \rightarrow MoO_2I + KCl + I_2 + H_2O$

66. $K_2SeO_3 + KI + HCl \rightarrow KCl + Se + I_2 + H_2O$

67. $SbCl_3 + I_2 + HCl \rightarrow SbCl_5 + HI$

68. $KIO_3 + KI + HCl \rightarrow KCl + I_2 + H_2O$

69. $HI + HClO_3 \rightarrow HCl + I_2 + H_2O$

70. $FeCl_2 + K_2Cr_2O_7 + HCl \rightarrow FeCl_3 + KCl + CrCl_3 + H_2O$

71. $Sb_2(SO_4)_3 + KMnO_4 + H_2O \rightarrow H_3SbO_4 + K_2SO_4 + MnSO_4 + H_2SO_4$

72. $Ba(BrO_3)_2 + H_3AsO_3 \rightarrow BaBr_2 + H_3AsO_4$

73. $Cr_2(SO_4)_3 + KMnO_4 + H_2O \rightarrow H_2CrO_4 + MnSO_4 + K_2SO_4 + H_2SO_4$

74. $NaAsO_2 + I_2 + H_2O \rightarrow NaAsO_3 + HI$

75. $HBO_2 + KIO_3 + KI \rightarrow KBO_2 + H_2O + I_2$

76. $Al_2(SO_4)_3 + KI + KIO_3 + H_2O \rightarrow Al(OH)_3 + K_2SO_4 + I_2$

77. $HAsO_2 + Ce(SO_4)_2 + H_2O \rightarrow H_3AsO_4 + Ce_2(SO_4)_3 + H_2SO_4$

78. $HNO_3 + H_2S \rightarrow S + NO + H_2O$

Nuclear Chemistry

Name _____

Date _____

Group I. Complete the following nuclear reactions.

1. $^{63}_{29}\text{Cu} + ^2_1\text{H} \rightarrow 2^1_0\text{n} +$ _____

2. $^{44}_{20}\text{Ca} + ^1_1\text{H} \rightarrow ^{44}_{21}\text{Sc} +$ _____

3. $^9_4\text{Be} + ^4_2\text{He} \rightarrow ^{12}_6\text{C} +$ _____

4. $^{31}_{15}\text{P} + ^2_1\text{H} \rightarrow ^{32}_{15}\text{P} +$ _____

5. $^{37}_{17}\text{Cl} +$ _____ $\rightarrow ^{35}_{16}\text{S} + ^4_2\text{He}$

6. $^2_1\text{H} + ^2_1\text{H} \rightarrow ^1_1\text{H} +$ _____

7. $^{11}_5\text{B} + ^4_2\text{He} \rightarrow ^{14}_7\text{N} +$ _____

8. $^{63}_{29}\text{Cu} + ^2_1\text{H} \rightarrow ^{64}_{30}\text{Zn} +$ _____

9. $^2_1\text{H} +$ _____ $\rightarrow ^1_1\text{H} + ^1_0\text{n}$

10. $^{31}_{15}\text{P} + ^1_1\text{H} \rightarrow ^{28}_{14}\text{Si} +$ _____

11. $^{239}_{94}\text{Pu} + ^4_2\text{He} \rightarrow ^1_0\text{n} +$ _____

12. $^{63}_{29}\text{Cu} + ^1_1\text{H} \rightarrow ^{38}_{17}\text{Cl} + ^1_0\text{n} +$ _____

(continued)

Name _____

Date _____

13. $^{63}_{29}\text{Cu} + ^{2}_{1}\text{H} \rightarrow ^{64}_{29}\text{Cu} + $ _____

14. $^{235}_{92}\text{U} + ^{1}_{0}\text{n} \rightarrow ^{95}_{42}\text{Mo} + 2^{1}_{0}\text{n} + $ _____

15. $^{6}_{3}\text{Li} + ^{2}_{1}\text{H} \rightarrow ^{7}_{4}\text{Be} + $ _____

16. $^{6}_{3}\text{Li} + ^{1}_{0}\text{n} \rightarrow ^{4}_{2}\text{He} + $ _____

17. $^{122}_{51}\text{Sb} + ^{1}_{0}\text{n} \rightarrow ^{122}_{51}\text{Sb} + $ _____

18. $^{214}_{82}\text{Pb} \rightarrow ^{0}_{-1}\text{e} + $ _____

19. $^{63}_{29}\text{Cu} + ^{2}_{1}\text{H} \rightarrow ^{3}_{1}\text{H} + $ _____

20. $^{14}_{7}\text{N} + $ _____ $\rightarrow ^{14}_{6}\text{C} + ^{1}_{1}\text{H}$

21. _____ $\rightarrow ^{237}_{93}\text{Np} + \alpha$

22. $^{2}_{1}\text{H} + ^{2}_{1}\text{H} \rightarrow ^{3}_{2}\text{He} + $ _____

23. $^{28}_{14}\text{Si} + ^{2}_{1}\text{H} \rightarrow ^{29}_{14}\text{Si} + $ _____

24. $^{9}_{4}\text{Be} + ^{4}_{2}\text{He} \rightarrow ^{6}_{3}\text{Li} + $ _____

25. $^{59}_{27}\text{Co} + ^{1}_{0}\text{n} \rightarrow ^{60}_{27}\text{Co} + $ _____

(continued)

Name _____

Date _____

26. $^{40}_{18}\text{Ar} + ^{2}_{1}\text{H} \rightarrow ^{41}_{18}\text{Ar} +$ _____

27. $^{6}_{3}\text{Li} + ^{2}_{1}\text{H} \rightarrow ^{1}_{0}\text{n} + ^{3}_{2}\text{He} +$ _____

28. $^{29}_{14}\text{Si} + ^{2}_{1}\text{H} \rightarrow ^{1}_{0}\text{n} +$ _____

29. $^{11}_{5}\text{B} + ^{1}_{1}\text{H} \rightarrow 3\text{X} \ (\text{X} = ?)$ _____

Group II. Write out in longhand (as in the previous examples) and complete the following nuclear reactions.

1. Li^6 (p, γ)

2. Na^{23} (α,) Mg^{26}

3. Cr^{54} (, n) Mn^{54}

4. C^{12} (p, γ)

5. Te^{125} (p, n)

6. B^{10} (, n) N^{13}

7. Al^{27} (n, p)

8. Li^7 (p, n)

(continued)

Chemistry Problems

Group III. Suggest a method by which each of the following isotopes can be produced, given the specified conditions (i.p. = incident particle; t = target nucleus).

1. $^{154}_{63}\text{Eu}$ $(t = {}^{152}_{62}\text{Sm})$

2. $^{18}_{9}\text{F}$ (i.p. = p)

3. $^{46}_{21}\text{Sc}$ (i.p. = n)

4. $^{95}_{41}\text{Nb}$ $(t = {}^{94}_{40}\text{Zr})$

5. $^{85}_{36}\text{Kr}$ (i.p. = d)

6. $^{142}_{59}\text{Pr}$ $(t = {}^{142}_{58}\text{Ce})$

7. $^{90}_{39}\text{Y}$ $(t = {}^{88}_{38}\text{Sr})$

8. $^{103}_{46}\text{Pd}$ (i.p. = α)

Group IV. In the following problems, an original sample of mass m_0 decays over time, t, to a mass of m_t. The half life of the isotope is given as $t_{1/2}$. In each problem, calculate the value of the missing variable.

1. $m_0 = 64$ g;   t = 4 half lives; $m_t = ?$

2. $m_0 = 32$ g;   t = 8.0 yr; $t_{1/2} = 2.0$ yr; $m_t = ?$

(continued)

                   Chemistry Problems

3. $m_0 = 128$ g; $t = 5$ d; $t_{1/2} = 24$ hr; $m_t = ?$

4. $m_0 = 50.0$ g; $t = 8.0$ min; $t_{1/2} = 1.6$ min; $m_t = ?$

5. $m_0 = 30.0$ g; $t = 9840$ yr; $t_{1/2} = 2460$ yr; $m_t = ?$

6. $m_0 = 400.0$ g; $m_t = 50.0$ g; $t_{1/2} = 13.6$ d; $t = ?$

7. $m_0 = 1.0000$ g; $m_t = 0.15625$ g; $t_{1/2} = 2325$ yr; $t = ?$

8. $m_t = 1.00$ g; $t_{1/2} = 10.0$ min; $t = 40.0$ min; $m_0 = ?$

9. $m_t = 2.5$ g; $t = 16.2$ yr; $t_{1/2} = 3.24$ yr; $m_0 = ?$

10. $m_0 = 64.0$ g; $m_t = 1.0$ g; $t = 18$ min; $t_{1/2} = ?$

Organic Chemistry

Name _____

Date _____

Group I. Identify the structural group in each of the following compounds and tell which organic family the compound belongs to.

1. $CH_3CH_2CH_3$

2. CH_3CH_2OH

3. $CH_3-C\overset{O}{\underset{H}{\diagup\!\!\diagup}}$

4. $CH_3CH(CH_3)CH_2CH_3$

5. $CH_3CH_2CH_2CHO$

6. CH_3OCH_3

7. $CH_3-\overset{}{\underset{\underset{O}{\|}}{C}}-CH_2CH_3$

8. $CH_2{=}CHCH_2CH_3$

9. $CH_3CHClCH_2CH_3$

10. $CH_3CH_2CH_2C\overset{O}{\underset{OH}{\diagup\!\!\diagup}}$

11. $CH_3C \equiv CH$

(continued)

Name _____

Date _____

Group II. **Write an expanded structural formula for each of the following compounds, and then name the compound.**

1. $CH_3CH_2CH_2CH_2CH_2CH_3$

2. $CH_3CH_2CH(CH_3)CH_2CH_2CH_2CH_3$

3. $CH_3CH(CH_3)C(CH_3)(C_2H_5)CH_2CH_2CH_3$

4. $CH_3CH_2CH_2CH_3$

5. $CH_3CH_2CH(C_2H_5)CH_2CH_3$

6. $CH_3CH_2CH(C_2H_5)CH_2CH_2CH(CH_3)CH_3$

7. $CH_3CH_2CH_2CH(CH_3)CH(CH_3)CH(CH_3)CH_3$

8. $CH_3CHBrCH_2CH(CH_3)CH_2CH_3$

9. $CH_3CH_2CH(CH_3)CH(CH_3)CH(C_2H_5)CH_2CH_3$

10. $CH_3CH_2CH_2CHClCHClCH_2CH_2CH(CH_3)CH_3$

11. $CH_2ClCHBrCH_2CH(CH_3)CH_2CH_2CHBrCH_3$

12. $CH_3CH(CH_3)CH_2CH(C_3H_7)CH(C_2H_5)CH_2CHClCH_3$

(continued)

© 1984, 1993, 2001 J. Weston Walch, Publisher

Chemistry Problems

13. $CH_3CH_2CH_2CH_2CH(C_2H_5)CH_2CHBrCH_2CH_3$

14. $CH_2ICH_2CH_2CHICH_2CHICH_2CH_3$

15. $CH_3CH_2CHFCH_2CH(CH_3)CHICH_2CH_3$

16. $CH_3CH(CH_3)CH_2CH(C_2H_5)CH_2CHICH_2CH_2CH_3$

17. $CH_2=CH_2$

18. $CH_3CH_2CH_2CH=CHCH_2CH_3$

19. $CH_3CH(CH_3)C(CH_3)=CHCH_2CH_3$

20. $CH_2=CHCH_2CH(CH_3)CHClCH_3$

21. $CH_3CH_2CHBrCH_2CH(CH_3)CH=CHCH_3$

22. $CH_3CH=CHCH_2CH(CH_3)CH(CH_3)CH_3$

23. $CH_3CH(CH_3)CH=C(C_2H_5)CH_2CH_2CH_2Cl$

24. $CH_3CH=CHC(CH_3)_2CH_2CHClC(CH_3)_2CH_2CH_3$

25. $CH_3CH=CHCH_2CCl_2CH_2CH_2Br$

(continued)

26. $CH_3CH_2CH{=}C(C_2H_5)CH_2CH_2CH_3$

27. $CH_3CHBrCH{=}CBrCH_2CH(CH_3)CH_3$

28. $CH_3CH_2CH{=}C(CH_3)CHBrCH_2CH_2Br$

29. $CH_3CH_2CH{=}CHCHBrCH_2CHClCH_2Cl$

30. $CH_3C{\equiv}CH$

31. $CH_3C{\equiv}CCH_3$

32. $CH_3CH_2C{\equiv}CCH_2Br$

33. $CH_3CH(CH_3)CH_2C{\equiv}CCH_3$

34. $CH_3CH_2CH_2C{\equiv}CCH(CH_3)CHBrCH_3$

35. $CH_3CH_2CH_2OH$

36. $CH_3CHOHCH_2CH_2CH_3$

37. $CH_3CH_2CHOHCH_3$

38. $CH_3CH_2CH_2CHOHCH_2CH_2CH_3$

(continued)

© 1984, 1993, 2001 J. Weston Walch, Publisher

Chemistry Problems

39. $CH_3CHOHCH_2CH_2CH_2CH_2CH_3$

40. $CH_3CH_2CHOHCHClCH_3$

41. $CH_3CH_2CH(CH_3)CH_2OH$

42. $CH_3CHOHCH(CH_3)CH_2CH_3$

43. $CH_3CH(CH_3)CH_2CH(CH_3)CHOHCH_3$

44. $CH_3C(CH_3)_2CH_2CHOHCH_2CH(CH_3)CH_3$

45. $CH_2OHCH_2CH_2CHClCH(CH_3)CH_3$

46. $CH_3CH(CH_3)CHOHCHClCH_2Cl$

47. $CH_3CH_2CH_2CHOHCH(CH_3)CH(CH_3)CH_3$

48. $CH_2ClCHOHCHClCH_3$

49. $CH_3CHOHCH_2CH_2C(C_2H_5)_2CH_2CH_3$

50. $CH_3CHOHCH_2CHClCH_2CHBrCH_2Cl$

51. $CH_2OHCH(CH_3)CH(C_2H_5)CH_2CH_2CH_3$

(continued)

Name _____

Date _____

52. $CH_2OHCH_2CH_2CH_2OH$

53. $CH_3CH_2CHOHCH_2CH_2CH_2OH$

54. CH_3CHO

55. $CH_3CH_2\overset{\displaystyle O}{\overset{\displaystyle \|}{C}}CH_2CH_2CH_3$

56. $CH_3CH_2CH_2CHO$

57. $CH_3CH(CH_3)CH_2CHO$

58. $CH_3CH_2C(CH_3)_2CH_2CHO$

59. $CH_3CH_2CHClCH_2CHO$

60. $C(CH_3)_3CH_2CH(CH_3)CHO$

61. $C(CH_3)_3CH_2CH_2CHO$

62. $CH_3CH_2\overset{\displaystyle O}{\overset{\displaystyle \|}{C}}CH_2CH_2CH_2CH_2CH_3$

63. $CH_3CH_2CH_2\overset{\displaystyle O}{\overset{\displaystyle \|}{C}}CH_2CH(CH_3)CH_3$

(continued)

© 1984, 1993, 2001 J. Weston Walch, Publisher

Chemistry Problems

Organic Chemistry *(continued)*

Name _____

Date _____

64. $CH_3C(CH_3)_2CH_2\overset{\overset{\textstyle O}{\|}}{C}CH_2CH_3$

65. $CH_3CHClCH_2CH_2CHClCH_2CHO$

66. $CH_3\overset{\overset{\textstyle O}{\|}}{C}CH_2CH_2CHICH_2CH(CH_3)_2$

67. $CH_3CHICH_2CH_2CHICHO$

68. $CH_3C(CH_3)_2CH_2CHI\overset{\overset{\textstyle O}{\|}}{C}CH_2CH_3$

69. $CH(CH_3)_2CH_2CH_2\overset{\overset{\textstyle O}{\|}}{C}CH_2CH_2CH_2Br$

70. $CH_3CH_2\overset{\overset{\textstyle O}{\|}}{C}CH_2CCl_2CH_2CH_2Cl$

71. $CH_3CH_2CH_2CH(CH_3)CH_2CHO$

72. $CH_3C(CH_3)_2CH_2CH_2\overset{\overset{\textstyle O}{\|}}{C}CH_2CH_2CH_2Cl$

73. $CH_3CH_2CHClCH_2C(CH_3)_2CH_2CHO$

74. CH_3OCH_3

75. $CH_3CH_2OCH_3$

(continued)

© 1984, 1993, 2001 J. Weston Walch, Publisher   129   *Chemistry Problems*

Name ______________________

Date ______________________

76. $CH_3CH_2CH_2OCH_3$

77. $CH_3OCH_2CH_2CH_2CH_2CH_3$

78. $CH_3CH_2OC(CH_3)_3$

79. $CH_3CH_2CH_2COOH$

80. $CH_3CH_2CH_2CH_2COOH$

81. $CH_3CH_2C(CH_3)_2COOH$

82. $CH_3CH(CH_3)CH_2COOH$

83. $HOOCCH_2CH_2COOH$

84. $CH_3CH_2CH(CH_3)CH_2CH_2COOH$

85. $CH_3CH(CH_3)CH_2CH(CH_3)COOH$

86. $CH_3CH_2CH_2CHClCH_2COOH$

87. $CH_3CH_2CHFCH_2CH_2COOH$

88. $CH_3CH_2CHNH_2COOH$

(continued)

Name ____________________

Date ____________________

89. $CH_3CH_2CHBrCOOH$

90. $CH_3CHBrCH_2CHBrCOOH$

91. $CH_3CH_2CH_2CHNH_2COOH$

92. $CH_3CHOHCH_2COOH$

93. $CH_3CH_2CH_2CH_2CHOHCOOH$

Group III. Draw structural formulas for each of the following compounds.

1. n-butane

2. 2-chloropentane

3. 3-methylhexane

4. n-heptane

5. 1, 2, 4-trichloropentane

6. 2, 3, 4-trimethylhexane

7. 2, 3, 3-trimethylheptane

8. 4-bromo-4-chloro-2-methyldecane

(continued)

Name ____________________

Date ____________________

9. 2, 3-dimethyl-3-hexene

10. 2-chloro-1-pentene

11. 2, 2, 3, 5-tetramethyl-3-octene

12. 1, 4, 5-tribromo-2, 2-dichloro-3-heptene

13. 3-chloro-2, 3-difluoro-1-butene

14. 1, 3-butadiene

15. ethanol

16. 2-propanol

17. 2-pentanol

18. 3, 3-dimethyl-2-hexanol

19. 1, 2, 3-trichloro-1-butanol

20. 3-ethyl-2, 2-dimethyl-1-hexanol

21. 2, 4, 4-trimethyl-2-heptanol

(continued)

22. 2-bromo-1, 3, 6, 7-tetrachloro-4, 4-dimethyl-1-decanol

23. 1, 3-butanediol

24. 2, 2-dibromo-3, 4-dimethyl-3-hexanol

25. 3-ethyl-2, 4-diiodo-2, 4, 5-trimethyl-1-nonanol

26. heptanal

27. butyraldehyde

28. methylpropyl ketone

29. diethyl ketone

30. 4-nonanone

31. 3-ethyl-2-methylhexanal

32. 3-bromo-2, 2-dichlorobutanal

33. 2-chloro-2, 4-dimethyl-3-hexanone

34. methanal

(continued)

Chemistry Problems

35. 2, 3, 3-trimethylpentanal

36. 4-chloro-4-iodo-3, 3-dimethyl-2-pentanone

37. 2, 3-dibromo-4, 6, 7-trichlorodecanal

38. 5, 7-diethyl-2, 3, 5-trimethyl-4-nonanone

39. butanoic acid

40. 1, 5-pentanedioic acid

41. α-aminopropionic acid

42. 2, 3-dichlorobutanoic acid

43. 3, 3-dimethylhexanoic acid

44. γ-hydroxyvaleric acid

45. 2, 3, 4-tribromodecanoic acid

46. 4, 4, 5-triiodo-2-methylhexanoic acid

47. α-hydroxybutanoic acid

(continued)

48. β-hydroxyvaleric acid

49. 4-chloro-2, 3-dimethylhexanoic acid

50. 3, 4-dibromo-2, 5, 7-trichlorodecanoic acid

51. 2, 2-diaminobutanoic acid

52. 5, 6-dichlorooctanoic acid

53. 2, 4-dihydroxypentanoic acid

Group IV. Write equations, using structural formulas, for the following organic reactions. Include catalysts or special conditions that may be necessary for each reaction to occur.

1. methane + chlorine (in four steps) →

2. propene + chlorine →

3. ethyl chloride + sodium hydroxide →

4. ethane + bromine →

5. 3-hexene + hydrogen chloride →

6. 2-butene + hydrogen →

(continued)

Name _____

Date _____

7. benzene + chlorine →

8. octane + oxygen (high temperature) →

9. 2-pentanone + hydrogen cyanide →

10. 1-butanol + sodium →

11. n-butane + bromine →

12. acetylene + hydrogen chloride →

13. propene (polymerized) →

14. benzene + bromine (in two steps) →

15. methanoic acid + 3-hexanol →

16. 1-pentene + hydrogen bromide →

17. n-dodecane + oxygen (high temperature) →

18. toluene + bromine →

19. 2-octene + butane →

(continued)

Name ______________________

Date ______________________

20. propane + chorine →

21. 1-butene + hydrogen →

22. 2-pentene + hydrogen cyanide →

23. 2-butanol + oxygen (mild conditions) →

24. 1-butene (polymerized) →

25. propyne + bromine →

26. n-pentane + bromine →

27. 2-pentanol + sodium →

28. 2-hexene + propane →

29. 2-pentene + bromine →

30. 3-hexanol + phosphorus trichloride →

31. n-octadecane + oxygen (high temperature) →

32. butene + ethane →

(continued)

Name _____

Date _____

33. pentanoic acid + n-butanol →

34. 3-hexanone + hydrogen cyanide →

35. 3-hexanol + oxygen (mild conditions) →

36. glyceryl palmitate distearate (hydrolyzed) →

37. ethanol + ethanoic acid →

38. alanine (α-aminopropionic acid) (polymerized) →

39. glyceryl trilaurate (hydrolyzed) →

40. ethanol + potassium →

41. 2-butanol + propanoic acid →

42. 1-butyne + hydrogen cyanide →

43. benzene (completely chlorinated) →

44. glycine (aminoacetic acid) (polymerized) →

45. 2-methylbutane + bromine →

(continued)

Chemistry Problems

Name ______________________

Date ______________________

46. glycine + alanine →

47. 2-hexanol + hydrogen chloride →

48. butanal + hydrogen cyanide →

49. butyric acid + butyl alcohol →

50. hexanal (oxidized) →

51. 2-methyl-1-pentene + bromine →

52. phenol + sodium hydroxide →

53. formaldehyde + ammonia →

54. glycerine + butyric acid →

55. phenol + formaldehyde →

56. pentanal + hydrogen chloride →

57. 2-hexene + hydrogen chloride →

58. benzene + sulfuric acid →

(continued)

Name ______________________

Date ______________________

59. t-butyl alcohol + phosphorus trichloride →

60. 3-pentanone + hydrogen →

61. methyl alcohol + potassium →

62. propanal + hydrogen →

63. hexane + bromine →

64. benzene + nitric acid (in two steps) →

65. n-octane + fluorine →

Quantum Mechanics

Name _____

Date _____

Group I. This series of questions deals with the wavelength (λ), frequency (f), and energy (E) of certain kinds of light. From the information given in each question, determine the missing variables.

1. $\lambda = 500.$ nm

2. $f = 6.0 \times 10^{14}$ Hz

3. $E = 4.5 \times 10^{-19}$ J

4. $\lambda = 850.$ nm

5. $f = 4.5 \times 10^{13}$ Hz

6. $E = 8.62 \times 10^{-18}$ J

7. $\lambda = 1250$ nm

8. $f = 8.22 \times 10^{15}$ Hz

9. $E = 1.22 \times 10^{-20}$ J

10. $\lambda = 355$ nm

11. $f = 3.362 \times 10^{14}$ Hz

12. $E = 3.604 \times 10^{-17}$ J

(continued)

Group II. The following questions concern electron transitions.

1. Calculate the total energy, in joules, required to ionize one mole of hydrogen atoms.

2. Find the values of E_n for the first three orbits of the hydrogen atom.

3. What energy is associated with a photon whose frequency is 4.3×10^{14} cycles per second?

4. What energy is emitted when an electron drops from the fourth to the second energy level in a hydrogen atom?

5. 10 Kcal/mole are emitted when an electron makes a particular transition from one orbit to another. What is the frequency of the photon that carries off this energy?

6. The energy absorbed by a particular atom during an electron transition from a lower to a higher level occurs as the absorption of red light of frequency 4.5×10^{14} cycles per second. What amount of energy is related to light of this frequency?

7. What is the total amount of energy absorbed by 1.0 mole of the substance in the case described in question 6?

8. An electron transition from a higher to a lower energy level in an atom results in a release of energy of 165.1 kJ/mol. What is the frequency of the photon that carries off this energy?

9. Express the radius of the third orbit of the hydrogen atom, in terms of the radius of its lowest energy level, r.

10. Using Einstein's expression for the mass equivalence of energy, $E = mc^2$, express the wavelength of an electron in terms of its mass.

Electrochemistry

Name _____

Date _____

1. How much silver metal will be plated out in an electrolytic cell containing Ag^+ when 0.50 mol of electrons passes through the cell?

2. How much zinc metal will 2.5 mol of electrons plate out of a zinc chloride solution?

3. A current of 5.0 A is passed through a solution of aluminum chloride for 15 minutes. How much aluminum metal is plated out during this period of time?

4. How many amperes of current are needed to plate out 10.0 g of silver per hour from a solution of silver nitrate?

5. Determine the standard potential for each of the following cells and tell which half cell is the cathode in each case.

   (a) Fe^{3+} (aq) | Fe^{2+} (aq); Ni^{2+} (aq) | Ni (s)

   (b) Cu^{2+} (aq) | Cu (s); Ag^+ (aq) | Ag (s)

   (c) Zn^{2+} (aq) | Zn (s); Ag^+ (aq) | Ag (s)

   (d) Cu^{2+} (aq) | Cu (s); Cd^{2+} (aq) | Cd (s)

   (e) Sn^{2+} (aq) | Sn (s); Br_2^0 (1) | Br (aq)

   (f) Al^{3+} (aq) | Al (s); Cu^{2+} (aq) | Cu (s)

   (g) Ni^{2+} (aq) | Ni (s); Cr^{3+} (aq) | Cr (s)

   (h) Cd^{2+} (aq) | Cd (s); Co^{2+} (aq) | Co (s)

Appendix A
Some Useful Physical Constants

Quantity	Symbol	Numerical Value
speed of light	c	2.9979246×10^8 m/s
elementary charge	e	$1.6021892 \times 10^{-19}$ C
Planck constant	h	6.626176×10^{-34} J/Hz
Avogadro number	N_A	6.022045×10^{23}/mol
rest mass of electron	m_e	9.109534×10^{-31} kg
rest mass of proton	m_p	$1.6726485 \times 10^{-27}$ kg
molar gas constant	R	$8.31451 \frac{\text{kPa} \cdot \text{dm}^3}{\text{mol} \cdot \text{K}}$

Appendix B
Units of Measure

Base Units in the SI System

Quantity	Unit	Symbol
length	meter	m
mass	kilogram	kg
time	second	s
temperature	kelvin	K
electric current	ampere	A
luminous intensity	candela	cd
amount of substance	mole	mol

Derived Units in the SI System

Quantity	Unit	Symbol
force	newton	N
pressure	pascal	Pa
energy	joule	J
power	watt	W
charge	coulomb	C
voltage	volt	V

(continued)

Other Common Units (not SI)

Quantity	Unit	Symbol
volume	liter	L
temperature	degree Celsius	°C

Metric Prefixes

Prefix	Symbol	Value
exa-	E	10^{18}
peta-	P	10^{15}
tera-	T	10^{12}
giga-	G	10^{9}
mega-	M	10^{6}
kilo-	k	10^{3}
hecto-	h	10^{2}
deka-	da	10^{1}
deci-	d	10^{-1}
centi-	c	10^{-2}
milli-	m	10^{-3}
micro-	μ	10^{-6}
nano-	n	10^{-9}
pico-	p	10^{-12}
femto-	f	10^{-15}
atto-	a	10^{-18}

Appendix C
Some English-Metric Conversion Factors

Length
1 cm = 0.3937 in
1 ft = 0.3048 m

Volume
1 L = 0.26417 U.S. gal
1 in^3 = 16.387 cm^3

Mass
1 kg = 2.2046 lb
1 oz = 28.3495 g

Temperature
°C = 5/9 (°F – 32)
°F = 9/5 (°C + 32)

Appendix D

Atomic Mass and Electron Configuration of the Elements

Element	Symbol	Atomic Number	Atomic Mass	Electron Configuration
actinium	Ac	89	(227)	Rn core $6d^17s^2$
aluminum	Al	13	26.98154	Ne core $3s^23p$
americium	Am	95	(243)	Rn core $5f^77s^2$
antimony	Sb	51	121.75	Kr core $4d^{10}5s^25p^3$
argon	Ar	18	39.948	$1s^22s^22p^63s^23p^6$
arsenic	As	33	74.9216	Ar core $3d^{10}4s^24p^3$
astatine	At	85	(210)	Xe core $4f^{14}5d^{10}6s^26p^5$
barium	Ba	56	137.33	Xe core $6s^2$
berkelium	Bk	97	(249)	Rn core $5f^86d^17s^2$
beryllium	Be	4	9.01218	He core $2s^2$
bismuth	Bi	83	208.9804	Xe core $4f^{14}5d^{10}6s^26p^3$
bohrium	Bh	107	(264)	Rn core $5f^{14}6d^57s^2$
boron	B	5	10.81	He core $2s^22p$
bromine	Br	35	79.904	Ar core $3d^{10}4s^24p^5$
cadmium	Cd	48	112.41	Kr core $4d^{10}5s^2$
calcium	Ca	20	40.08	Ar core $4s^2$
californium	Cf	98	(251)	Rn core $5f^{10}7s^2$
carbon	C	6	12.011	He core $2s^22p^2$
cerium	Ce	58	140.12	Xe core $4f^15d^16s^2$
cesium	Cs	55	132.9054	Xe core $6s$
chlorine	Cl	17	35.453	Ne core $3s^23p^5$
chromium	Cr	24	51.996	Ar core $3d^54s^1$
cobalt	Co	27	58.9332	Ar core $3d^74s^2$
copper	Cu	29	63.546	Ar core $3d^{10}4s^1$
curium	Cm	96	(247)	Rn core $5f^76d7s^2$
dubnium	Db	105	(262)	Rn core $5f^{14}6d^37s^2$
dysprosium	Dy	66	162.50	Xe core $4f^{10}6s^2$
einsteinium	Es	99	(254)	Rn core $5f^{11}7s^2$
erbium	Er	68	167.26	Xe core $4f^{12}6s^2$
europium	Eu	63	151.96	Xe core $4f^76s^2$
fermium	Fm	100	(253)	Rn core $5f^{12}7s^2$
fluorine	F	9	18.998103	He core $2s^22p^5$

(continued)

Chemistry Problems

Element	Symbol	Atomic Number	Atomic Mass	Electron Configuration
francium	Fr	87	(223)	Rn core $7s^1$
gadolinium	Gd	64	157.25	Xe core $4f^75d^16s^2$
gallium	Ga	31	69.72	Ar core $3d^{10}4s^24p$
germanium	Ge	32	72.59	Ar core $3d^{10}4s^24p^2$
gold	Au	79	196.9665	Xe core $4f^{14}5d^{10}6s^1$
hafnium	Hf	72	178.49	Xe core $4f^{14}5d^26s^2$
hassium	Hs	108	(269)	Rn core $5f^{14}6d^67s^2$
helium	He	2	4.00260	$1s^2$
holmium	Ho	67	164.9304	Xe core $4f^{11}6s^2$
hydrogen	H	1	1.0079	$1s$
indium	In	49	114.82	Kr core $4d^{10}5s^25p^1$
iodine	I	53	126.9045	Kr core $4d^{10}5s^25p^5$
iridium	Ir	77	192.22	Xe core $4f^{14}5d^76s^2$
iron	Fe	26	55.847	Ar core $3d^64s^2$
krypton	Kr	36	83.80	$1s^22s^22p^63s^23p^63d^{10}4s^24p^6$
lanthanum	La	57	138.9055	Xe core $5d^16s^2$
lawrencium	Lr	103	(257)	Rn core $5f^{14}6d7s^2$
lead	Pb	82	207.2	Xe core $4f^{14}5d^{10}6s^26p^2$
lithium	Li	3	6.941	He core $2s$
lutetium	Lu	71	174.97	Xe core $4f^{14}5d^16s^2$
magnesium	Mg	12	24.305	Ne core $3s^2$
manganese	Mn	25	54.94	Ar core $3d^54s^2$
meitnerium	Mt	109	(268)	Rn core $5f^{14}6d^77s^2$
mendelevium	Md	101	(256)	Rn core $5f^{13}7s^2$
mercury	Hg	80	200.59	Xe core $4f^{14}5d^{10}6s^2$
molybdenum	Mo	42	95.94	Kr core $4d^55s^1$
neodymium	Nd	60	144.24	Xe core $4f^46s^2$
neon	Ne	10	20.179	$1s^22s^22p^6$
neptunium	Np	93	237.0482	Rn core $5f^46d7s^2$
nickel	Ni	28	58.70	Ar core $3d^84s^2$
niobium	Nb	41	92.9064	Kr core $4d^45s^1$
nitrogen	N	7	14.0067	He core $2s^22p^3$
nobelium	No	102	(254)	Rn core $5f^{14}7s^2$
osmium	Os	76	190.2	Xe core $4f^{14}5d^66s^2$
oxygen	O	8	15.9994	He core $2s^22p^4$
palladium	Pd	46	106.4	Kr core $4d^{10}$

(continued)

Element	Symbol	Atomic Number	Atomic Mass	Electron Configuration
phosphorus	P	15	30.97376	Ne core $3s^23p^3$
platinum	Pt	78	195.09	Xe core $4f^{14}5d^96s^1$
plutonium	Pu	94	(242)	Rn core $5f^67s^2$
polonium	Po	84	(210)	Xe core $4f^{14}5d^{10}6s^26p^4$
potassium	K	19	39.0983	Ar core $4s$
praesodymium	Pr	59	140.9077	Xe core $4f^36s^2$
promethium	Pm	61	(145)	Xe core $4f^56s^2$
protactinium	Pa	91	231.0359	Rn core $5f^26d7s^2$
radium	Ra	88	226.0254	Rn core $7s^2$
radon	Rn	86	(222)	$1s^22s^22p^63s^23p^63d^{10}4s^24p^6$ $4d^{10}4f^{14}5s^25p^65d^{10}6s^26p^6$
rhenium	Re	75	186.207	Xe core $4f^{14}5d^56s^2$
rhodium	Rh	45	102.9055	Kr core $4d^85s^1$
rubidium	Rb	37	85.4678	Kr core $5s$
ruthenium	Ru	44	101.07	Kr core $4d^75s^1$
rutherfordium	Rf	104	(261)	Rn core $5f^{14}6d^27s^2$
samarium	Sm	62	150.4	Xe core $4f^66s^2$
scandium	Sc	21	44.9559	Ar core $3d^14s^2$
seaborgium	Sg	106	(263)	Rn core $5f^{14}6d^27s^2$
selenium	Se	34	78.96	Ar core $3d^{10}4s^24p^4$
silicon	Si	14	28.0855	Ne core $3s^23p^2$
silver	Ag	47	107.868	Kr core $4d^{10}5s^1$
sodium	Na	11	22.98977	Ne core $3s$
strontium	Sr	38	87.62	Kr core $5s^2$
sulfur	S	16	32.06	Ne core $3s^23p^4$
tantalum	Ta	73	180.9479	Xe core $4f^{14}5d^36s^2$
technetium	Tc	43	98.9062	Kr core $4d^55s^2$
tellurium	Te	52	127.60	Kr core $4d^{10}5s^25p^4$
terbium	Tb	65	158.9254	Xe core $4f^96s^2$
thallium	Tl	81	204.37	Xe core $4f^{14}5d^{10}6s^26p^1$
thorium	Th	90	232.0381	Rn core $6d^27s^2$
thulium	Tm	69	168.9342	Xe core $4f^{13}6s^2$
tin	Sn	50	118.69	Kr core $4d^{10}5s^25p^2$
titanium	Ti	22	47.90	Ar core $3d^24s^2$
tungsten	W	74	183.85	Xe core $4f^{14}5d^46s^2$
ununnilium	Uun	110	(269)	Rn core $5f^{14}6d^97s^1$

(continued)

Element	Symbol	Atomic Number	Atomic Mass	Electron Configuration
unununium	Uuu	111	(272)	Rn core $5f^{14}6d^{10}7s^1$
ununbium	Uub	112	(277)	Rn core $5f^{14}6d^{10}7s^2$
ununquadium	Uuq	114	(285)	Rn core $5f^{14}6d^{10}7s^27p^2$
ununhexium	Uuh	116	(289)	Rn core $5f^{14}6d^{10}7s^27p^4$
ununoctium	Uuo	118	(293)	Rn core $5f^{14}6d^{10}7s^27p^6$
uranium	U	92	238.029	Rn core $5f^36d7s^2$
vanadium	V	23	50.9444	Ar core $3d^34s^2$
xenon	Xe	54	131.30	$1s^22s^22p^63s^23p^63d^{10}4s^24p^6$ $4d^{10}5s^25p^6$
ytterbium	Yb	70	173.04	Xe core $4f^{14}6s^2$
yttrium	Y	39	88.9059	Kr core $4d^15s^2$
zinc	Zn	30	65.38	Ar core $3d^{10}4s^2$
zirconium	Zr	40	91.22	Kr core $4d^25s^2$

Atomic mass is based on carbon 12. Values in parentheses are the mass numbers of the most stable or best known isotopes of the elements.

Appendix E
Density of Some Common Materials

Substance	Density
air	1.293 g/dm^3
alcohol (ethyl)	0.79 g/cm^3
aluminum	2.70 g/cm^3
brass	$8.4 – 8.7$ g/cm^3
carbon (graphite)	2.25 g/cm^3
carbon dioxide	1.997 g/dm^3
copper	8.89 g/cm^3
ether	0.74 g/cm^3
gasoline	0.68 g/cm^3
glass	$2.4 – 2.8$ g/cm^3
gold	19.3 g/cm^3
hydrogen	0.090 g/dm^3

(continued)

Substance	Density
ice	0.917 g/cm³
iron (wrought)	7.85 g/cm³
lead	11.3 g/cm³
mercury	13.6 g/cm³
nitrogen	1.25 g/dm³
oxygen	1.43 g/dm³
silver	10.5 g/cm³
steel	7.80 g/cm³
water (20 °C)	0.998 g/cm³
water (4 °C)	1.000 g/cm³
zinc	7.1 g/cm³

Appendix F
Electronegativities of the Elements

H 2.1																	
Li 1.0	Be 1.5											B 2.0	C 2.5	N 3.1	O 3.5	F 4.1	
Na 1.0	Mg 1.3											Al 1.5	Si 1.8	P 2.1	S 2.4	Cl 2.9	
K 0.9	Ca 1.1	Sc 1.2	Ti 1.3	V 1.5	Cr 1.6	Mn 1.6	Fe 1.7	Co 1.7	Ni 1.8	Cu 1.8	Zn 1.7	Ga 1.8	Ge 2.0	As 2.2	Se 2.5	Br 2.8	
Rb 0.9	Sr 1.0	Y 1.1	Zr 1.2	Nb 1.3	Mo 1.3	Tc 1.4	Ru 1.4	Rh 1.5	Pd 1.4	Ag 1.4	Cd 1.5	In 1.5	Sn 1.7	Sb 1.8	Te 2.0	I 2.2	
Cs 0.9	Ba 0.9	La 1.1	Hf 1.2	Ta 1.4	W 1.4	Re 1.5	Os 1.5	Ir 1.6	Pt 1.5	Au 1.4	Hg 1.5	Tl 1.5	Pb 1.6	Bi 1.7	Po 1.8	At 2.0	
Fr 0.9	Ra 0.9	Ac 1.0	Lanthanides: 1.0–1.2														
			Actinides: 1.0–1.2														

For practice problems, assume that differences in electronegativities of less than 0.4 between two bonding atoms result in a non-polar covalent bond, differences between 0.4 and 1.7 result in a polar covalent bond, and differences greater than 1.7 result in an ionic bond.

Appendix G

Heats of Formation

Compound	Heat of formation (in kilojoules per mole)	Compound	Heat of formation (in kilojoules per mole)
$AlCl_3$	695.4	P_2O_5	1506.2
$NH_3(g)$	46.0	KBr	392.0
NH_4Cl	315.5	KCl	436.0
$BaCl_2$	860.2	$KClO_3$	391.2
BaO	558.1	KI	327.6
$CaCO_3$	1206.7	KNO_3	492.9
$CaCl_2$	795.0	KOH	425.9
CaO	635.5	AgCl	127.2
$CH_4(g)$	74.9	$AgNO_3$	123.0
$CO_2(g)$	393.3	NaBr	359.8
CO(g)	110.5	NaCl	410.9
$CCl_4(l)$	138.9	NaF	569.0
$CuCl_2$	218.8	$SO_2(g)$	297.1
CuO	155.2	$SO_3(l)$	431.0
Cu_2O	166.5	$SnCl_2$	349.8
$CuSO_4$	769.9	$H_2O(l)$	285.8
HBr(g)	36.4	$H_2O_2(l)$	187.4
HCl(g)	92.5	$H_2S(g)$	20.1
HF(g)	268.6	$H_2SO_4(l)$	3823.8
HI(g)	−25.9	$FeCl_3$	405.0
$HNO_3(l)$	173.2	FeS	95.0
PbO	218.0	$SnCl_4(l)$	515.2
$MgCl_2$	641.8	$ZnCl_2$	415.9
HgO	90.8	$Zn(NO_3)_2$	481.6
NO(g)	−90.4	ZnO	348.1
N2O(g)	81.6	$Zn(OH)_2$	642.2
NO_2	−33.9	ZnS	202.9

Appendix H

Specific Heat, Heat of Fusion, and Heat of Vaporization of Some Selected Substances

Substance	Specific Heat (J/g • °C)	Heat of Fusion (J/g)	Heat of Vaporization (J/g)
alcohol (ethyl)	2.51	109	879
aluminum	0.900	395	10,800
benzene	1.42	126	548
chloroform	0.971	79.5	264
copper	0.385	205	5310
ether (ethyl)	2.34	100	368
ice	2.09	334	2260
iron	0.460	267	7450
lead	0.130	24.7	946
magnesium	1.02	372	6070
mercury	0.138	11.3	285
steam	2.02	334	2260
water	4.184	334	2260

Appendix I

Vapor Pressure of Water
(in millimeters of mercury)

temperature (°C)	pressure (kPa)	temperature (°C)	pressure (kPa)	temperature (°C)	pressure (kPa)
1	0.653	5	0.866	9	1.15
2	0.706	6	0.933	10	1.23
3	0.760	7	1.00	11	1.31
4	0.813	8	1.07	12	1.40

(continued)

Chemistry Problems

temperature (°C)	pressure (kPa)	temperature (°C)	pressure (kPa)	temperature (°C)	pressure (kPa)
13	1.49	29	4.00	45	9.58
14	1.60	30	4.24	46	10.1
15	1.71	31	4.49	47	10.6
16	1.81	32	4.76	48	11.2
17	1.93	33	5.03	49	11.7
18	2.07	34	5.32	50	12.3
19	2.20	35	5.62	55	15.7
20	2.33	36	5.94	60	19.9
21	2.48	37	6.28	65	25.0
22	2.64	38	6.62	70	31.1
23	2.81	39	6.98	75	38.5
24	2.99	40	7.37	80	47.3
25	3.17	41	7.77	85	57.8
26	3.36	42	8.20	90	70.1
27	3.56	43	8.64	95	84.5
28	3.77	44	9.10	100	101.3

Appendix J

Boiling Point Elevation and Freezing Point Depression Constants

Substance	Normal Boiling Point (°C)	Boiling Point Elevation (°C)	Normal Freezing Point (°C)	Freezing Point Depression (°C)
acetic acid	118.1	2.53	16.6	3.90
benzene	80.1	2.53	5.5	5.12
camphor	207.4	5.61	179.5	37.7
carbon tetrachloride	76.8	4.48	–23.0	29.8
chloroform	61.2	3.62	–63.5	4.68

(continued)

Chemistry Problems

Substance	Normal Boiling Point (°C)	Boiling Point Elevation (°C)	Normal Freezing Point (°C)	Freezing Point Depression (°C)
cyclohexane	80.7	2.75	6.5	6.54
1,2-dibromoethane	131.7	6.61	10	12.5
ethyl alcohol	78.4	1.16	–114.6	1.99
naphthalene	218.0	5.80	80.2	6.94
water	100.0	0.52	0.0	1.86

Appendix K
Activity Series of the Elements

Metals		Non-metals
lithium	cobalt	fluorine
rubidium	nickel	chlorine
potassium	tin	bromine
sodium	lead	iodine
strontium	HYDROGEN	sulfur
barium	antimony	
calcium	bismuth	
magnesium	arsenic	
aluminum	copper	
manganese	mercury	
zinc	silver	
chromium	platinum	
iron	gold	
cadmium		

Appendix L

Table of Solubilities

Note: All common compounds containing sodium or potassium or the ammonium, chlorate, or nitrate radical are soluble in water and are not listed in the following table.

S = soluble in water

P = partially soluble in water

I = insoluble in water and in dilute acids

A = insoluble in water, soluble in dilute acids

a = partially soluble in dilute acids; insoluble in water

X = does not exist or decomposes in water

	acetate	bromide	carbonate	chloride	chromate	hydroxide	iodide	oxide	phosphate	silicate	sulfate	sulfide
aluminum	S	S	X	S	X	A	S	a	A	I	S	X
barium	S	S	P	S	A	S	S	S	A	S	a	X
calcium	S	S	P	S	S	S	S	P	P	P	P	P
cupric	S	S	X	S	X	A	X	A	A	A	S	A
ferrous	S	S	P	S	X	A	S	A	A	X	S	A
ferric	S	S	X	S	A	A	S	A	P	X	P	X
lead	S	S	A	S	A	P	S	P	A	A	P	A
magnesium	S	S	P	S	S	A	S	A	P	A	S	X
manganese	S	S	P	S	X	A	S	A	P	I	S	A
mercurous	P	A	A	a	P	X	A	A	A	X	P	I
mercuric	S	S	X	S	P	A	P	P	A	X	X	I
silver	P	a	A	a	P	X	I	P	A	X	P	A
stannous	X	S	X	S	A	A	S	A	A	X	S	A
stannic	S	S	X	X	S	P	X	A	X	X	S	A
strontium	S	S	P	S	P	S	S	S	A	A	P	S
zinc	S	S	P	S	P	A	S	P	A	A	S	A

Appendix M

Half Lives of Some Common Radioactive Isotopes

Isotope	Half Life
^{3}H	12.3 y
^{14}C	5730 y
^{15}O	124 s
^{24}Na	15.0 h
^{32}P	14.3 d
^{40}K	1.28×10^9 y
^{51}Cr	27.8 d
^{55}Fe	2.6 y
^{60}Co	5.26 y
^{85}Sr	70 m
^{99m}Tc	6.0 h
^{131}I	8.1 d
^{198}Au	2.7 d
^{226}Ra	1600 y
^{234}Th	24 d
^{235}U	7.1×10^8 y
^{238}U	4.5×10^9 y
^{239}Np	2.35 d
^{239}Pu	24,400 y

Appendix N

Some Standard Reduction Potentials

(at 25°C and 1 mol conc)

Electrode	Half Reaction	E°
$Li^+ \mid Li$	$Li^+ + e^- \rightarrow Li$	-3.05
$K^+ \mid K$	$K^+ + e^- \rightarrow K$	-2.93
$Ca^{2+} \mid Ca$	$Ca^{2+} + 2\,e^- \rightarrow Ca$	-2.87
$Na^+ \mid Na$	$Na^+ + e^- \rightarrow Na$	-2.71
$Mg^{2+} \mid Mg$	$Mg^{2+} + 2\,e^- \rightarrow Mg$	-2.37
$Al^{3+} \mid Al$	$Al^{3+} + 3\,e^- \rightarrow Al$	-1.66
$Zn^{2+} \mid Zn$	$Zn^{2+} + 2\,e^- \rightarrow Zn$	-0.76
$Cr^{3+} \mid Cr$	$Cr^{3+} + 3e^- \rightarrow Cr$	-0.74
$Fe^{2+} \mid Fe$	$Fe^{2+} + 2\,e^- \rightarrow Fe$	-0.44
$Cd^{2+} \mid Cd$	$Cd^{2+} + 2\,e^- \rightarrow Cd$	-0.40
$Co^{2+} \mid Co$	$Co^{2+} + 2\,e^- \rightarrow Co$	-0.28
$Ni^{2+} \mid Ni$	$Ni^{2+} + 2\,e^- \rightarrow Ni$	-0.25
$Sn^{2+} \mid Sn$	$Sn^{2+} + 2\,e^- \rightarrow Sn$	-0.14
$Pb^{2+} \mid Pb$	$Pb^{2+} + 2\,e^- \rightarrow Pb$	-0.13
$Fe^{3+} \mid Fe$	$Fe^{3+} + 2\,e^- \rightarrow Fe$	-0.036
$H^+ \mid H_2$	$2H^+ + 2\,e^- \rightarrow H_2$	-0.00
$Cu^{2+} \mid Cu$	$Cu^{2+} + 2\,e^- \rightarrow Cu$	$+0.34$
$Cu^+ \mid Cu$	$Cu^+ + e^- \rightarrow Cu$	$+0.52$
$Fe^{3+} \mid Fe^{2+}$	$Fe^{3+} + e^- \rightarrow Fe^{3+}$	$+0.77$
$Hg_2^{2+} \mid Hg$	$Hg_2^{2+} + 2\,e^- \rightarrow 2\,Hg$	$+0.34$
$Ag^+ \mid Ag$	$Ag^+ + e^- \rightarrow Ag$	$+0.80$
$Hg_2^{2+} \mid Hg$	$Hg_2^{2+} + 2\,e^- \rightarrow Hg$	$+0.85$
$Br_2 \mid Br^-$	$Br_2 + 2\,e^- \rightarrow Br^-$	$+1.07$
$Cl_2 \mid Cl^-$	$Cl_2 + 2\,e^- \rightarrow Cl^-$	$+1.36$
$F_2 \mid F^-$	$F_2 + 2\,e^- \rightarrow F^-$	$+2.87$

Appendix O

Some Useful, Chemistry-Related Web Sites

The American Chemical Society
http://www.acs.org/

The National Science Teachers Association
http://www.nsta.org/

Chem 101
http://library.thinkquest.org/3310/

Tanner's General Chemistry
http://www.tannerm.com/

Atomic Structure Timeline
http://www.watertown.k12.wi.us/hs/teachers/buescher/atomtime.asp

The History of Chemistry
http://www.chemistrycoach.com/history_of_chemistry.htm#

About.com for Chemistry
http://chemistry.about.com/science/chemistry/

Chemdex: The directory of chemistry on the World Wide Web
http://www.chemdex.org/index.html

MSDS Resource from Cornell University
http://msds.pdc.cornell.edu/msdssrch.asp

A Visual Interpretation of the Table of Elements
http://www.chemsoc.org/viselements/

Yahoo! Chemistry
http://dir.yahoo.com/Science/Chemistry

The Alchemist
http://www.chemweb.com/alchem/2000/homepage/hp_current.html

Chemistry News
http://www.chemnews.com/

The Chemistry General Index
http://macedonia.chem.demokritos.gr/chemistry/

General Chemistry Online
http://antoine.frostburg.edu/chem/senese/101/

Martindale's Reference Desk: Chemistry Center
http://www-sci.lib.uci.edu/HSG/GradChemistry.html

Appendix P

Calculator Help for Scientific Notation and Logarithms

I. Scientific notation is generally used to make long numbers more manageable. The number is expressed as a decimal number between 1 and 10 multiplied by a power of 10. For example, the number 4 600 000 000 000 can be written as 4.6×10^{12}. The number 0.000 000 000 000 000 000 35 can be written as 3.5×10^{-19}. This is because the number 10 and the powers of 10 can be factored out of a longer number. For example:

$10 \times 10 = 10^2$ and $10 \times 10 \times 10 = 10^3$

$3000 = 3 \times 1000$ or $3 \times 10 \times 10 \times 10$ or 3×10^3

The same is true for decimals:

0.1 is 1/10

0.01 is $1/(10 \times 10)$ or $1/100$ or $1/10^2$

0.001 is $1/(10 \times 10 \times 10)$ or $1/1000$ or $1/10^3$

A negative exponent indicates that the base 10 is the denominator of a fraction whose numerator is 1.

$1/10^3$ is equal to 1×10^{-3}.

A number like 0.000 000 000 028 is numerically equivalent to $2.8 \times 1/10^{11}$ or 2.8×10^{-11}.

The goal of scientific notation is to convert a very large or very small number into a more manageable form. Some examples of numbers expressed in scientific notation are as follows:

4.54×10^{13}

2.33×10^{-6}

9×10^{18}

$2.309\ 303\ 987\ 783 \times 10^{68}$

The first digit in such numbers is always in the ones place and has a value of 1–9. The number zero is never used as the first digit. Numbers larger than nine are never used for the first digit, either. For example:

WRONG: 321.4×10^6 WRONG: 0.421×10^6
CORRECT: 3.214×10^8 CORRECT: 4.21×10^5

On a scientific calculator, the button that best allows you to enter numbers in scientific notation is the exponential button. This button is usually labeled "exp," "e," "ee," "EXP," "E," or "EE." The use of these buttons reduces the risk of the following:

1. Putting the number 10 in more times than necessary.

(continued)

2. Applying the exponent to a number other than 10.
   For example, the correct way to enter the number 3.2×10^{21} would be to type in
   " 3 . 2 EXP 2 1 ". Many students will type in " 3 . 2 x^y 2 1 " or " 3 . 2 ^ 2 1 ". Either
   of these last two methods will give the answer $4.056\ 481\ 921 \times 10^{10}$ (4.1×10^{10} in
   significant figures), when the correct answer is 32 followed by 20 zeros.

II. A logarithm is the power to which a base number must be raised in order to produce
a given number. For example, the log of 10 000 is 4; this is because 10 is raised to the
fourth power to make it equal to 10 000 ($10^4 = 10\ 000$). Logs also work for decimal
values. $0.000\ 001 = 10^{-6}$. Therefore, the log of 0.000 001 is –6.

Chemists use the log function with problems involving pH. The common formula for
calculating pH is:

$pH = -\log [H_3O^+]$ where H_3O^+ is the molar concentration of H_3O^+ ions in solution.

This formula is used in one of two ways. Either the concentration of H_3O^+ is known and
the pH value is to be calculated or the pH value is given and the concentration is to
be calculated.

When the pH is to be evaluated, the use of a scientific calculator is preferable to log
tables or estimating.

For example: What is the pH of a solution that is 0.000 32 M HCl?

In the calculator, you would enter " – log 0. 0 0 0 3 2 = ". The correct answer is
3.494 850 022, which should be rounded to 3.5 according to significant figures.

If you are calculating the concentration from known pH, the formula must be
"reversed" according to algebraic rules. This requires finding the antilog of both sides
and then multiplying both sides by –1.

Antilog (pH) = antilog ($-\log [H_3O^+]$) or

$10^{-pH} = [H_3O^+]$

To use this equation, most calculators have a 10^x button or the log button is used with
a shift, 2nd, or inverse button. (Note that the x in 10^x is **negative** pH.)

For example: What is the molar concentration of a solution that has a pH of 4.4?

In the calculator a student would enter "10^x – 4 . 4" and the answer would be 4.0×10^{-5}.

Answer Key

Note: Student answers may differ slightly from those given here, depending on the atomic weight table used and the method of rounding suggested by the teacher.

Exponential Notation (pp. 1–3)

Group I.

1. 1×10^4
2. 1×10^{-4}
3. 1×10^{10}
4. 5×10^4
5. 2×10^9
6. 4×10^{-7}
7. 3×10^{-4}
8. 7.9×10^5
9. 2.6×10^{13}
10. 4.5×10^{-8}
11. 4.8×10^7
12. 1.3×10^{-18}
13. 1.75×10^{-2}
14. 4.62×10^{-5}
15. 3×10^6
16. 9.2×10^6
17. 3.71×10^2
18. 9.00×10^{-3}
19. 6.2×10^{-1}
20. 1.97×10^{-4}
21. 3.285×10^5
22. 7.645×10^4
23. 9.410×10^{-1}
24. 3.005×10^3
25. 7.05×10^{-4}
26. 8×10^{10}
27. 6.24×10^3
28. 1.75×10^7
29. 5.302×10^{-2}
30. 9.15×10^{-12}
31. 2.8×10^1
32. 7.480×10^0
33. 2.805×10^8
34. 1.342×10^2
35. 8.9×10^{-5}

Group II.

1. 100 000
2. 0.000 000 000 001
3. 400
4. 0.000 005
5. 0.032
6. 0.795
7. 685 400 000 000 000
8. 1430
9. 0.000 906 5
10. 4 300
11. 442 000 000
12. 0.003 08
13. 16.9
14. 663 000
15. 0.000 004 00

Group III.*

1. 24
2. 1×10^{-5}
3. 20
4. 4.58×10^3
5. 1.35×10^{10}
6. 3×10^{-2}
7. 8×10^{-6}
8. 8×10^{-2}
9. 7.9×10^9
10. 1.47×10^5
11. 5×10^{-3}
12. 48
13. 8×10^{-7}
14. 3×10^3
15. 4×10^{-2}
16. 8.2×10^3
17. 2×10^{-3}
18. 2×10^{-4}

*In addition to those given here, equivalent correct answers are possible.

(continued)

19. 1.8×10^{-5}

20. 9×10^{-10}

21. 8×10^3

22. 5×10^{11}

23. 3.839×10^4

24. .24

25. 2×10^7

26. 5.2×10^{-2}

27. 4.48×10^3

28. 2.4×10^{-8}

29. 2×10^{-4}

30. 1.12×10^7

31. 6.1×10^{-2}

32. 8×10^{-10}

33. 4×10^6

34. 4.883×10^{-6}

35. 64

36. 2.8×10^{10}

37. 3.2

38. .01

39. 2×10^{-8}

40. 5×10^{-7}

41. 2×10^4

42. 4.5×10^6

43. 5×10^{-5}

44. 8×10^{-9}

45. 3×10^4

46. 2×10^5

47. 7×10^{-9}

48. 1.33×10^{14}

49. 5×10^{12}

50. 6.66×10^{-12}

Significant Figures (pp. 4–8)

Group I.

1. 4
2. 5
3. 3
4. 4
5. 3

6. 4
7. 5
8. 3
9. 3
10. 3

11. 2, 3, or 4
12. 5
13. 3
14. 3
15. 7

16. 5
17. 4
18. 4
19. 3
20. 4

Group II.

1. 20. cm
2. 0.29 cm
3. 14.383 g
4. 173.6 g
5. 506.7 cm

6. 246.52 cm
7. 0.2704 g
8. 3.89×10^{-4} cm
9. 270.6 cm^3
10. 10.8×10^2 g

Group III.

1. 17.74 cm
2. 245 g
3. 116.7 cm^3

4. 22.6 g
5. 10.914 cm^3

Group IV.

1. 11.6 cm^2
2. 110 cm^2
3. 3.7 m^2
4. 36.6 mm^2
5. 2560 dm^2

6. 20 cm^2
7. 20 cm^2
8. 25 cm^2
9. $10. \times 10^5$ m^2
10. 11×10^{-7} cm^2

11. 27 cm^2
12. 600. m^2
13. 850 km^2
14. 46×10^{-14} m^2
15. 4.21×10^7 m^2

Group V.

1. 1.914 cm
2. 18.82 m
3. 570. km
4. 7.1 mm
5. 0.4891 cm

6. 2.029 m
7. 0.223 m
8. 10.6 km
9. 1.036×10^2 mm
10. 1.3×10^{-5} mm

Group VI.

1. 0.064 cm^3
2. 137 cm^3
3. 19.9 g

4. 55.479 g
5. 7.329×10^{16} cm^3

Group VII.

1. 0.0109 g/cm^3; –7.37%
2. –0.02 g; –0.51%
3. 1.3°C; 0.75%
4. –1 kg; –1.23%
5. –0.0964 g; –90.77%
6. 0.5 cm^3; 0.5%
7. –0.69 dynes/cm; –0.96%
8. –0.001 3 g; –0.0232%
9. –0.000 9 N; –0.6598%
10. +9.3% error in first case; –9.3% error in the second. The errors have the same magnitude, but different directions.

11. 1.511 2 g/cm^3; 1.367 2 g/cm^3
12. 86.670 g; 82.442 g
13. Minimum value: 865.4 cm
    Maximum value: 866.2 cm
14. 0.03 g
15. 1269
16. 8
17. 50.0 ± 2.25 cm^2
18. D = 0.1823 ± 0.0006 g/cm^3
19. (a) 18.514 g   (b) 0.002 g
20. Heat absorbed = 23,220 ± 193 cal

Dimensional Analysis (p. 9)

1. 0.14 m
2. 31 000 mg
3. 0.116 5 km
4. 0.002 859 km
5. 639.4 cm
6. 0.84 cg
7. 0.147 km

8. 470 000 cg
9. 0.138 4 g
10. 65 500 m
11. 236 cm
12. 0.273 days
13. 13 600 cm^3
14. 5.72 m/s

15. 5.8 cm/s
16. 0.969 m/s
17. 1.47 cg/mm^3
18. 63 cm^3/day
19. 8.05×10^5 kg/dm^3
20. 342 g/cm^2

Metric System (pp. 10–17)

Group I.

1. 73 000 mg
2. 16 530 mg
3. 2.38 mg
4. 10.6 mg
5. 420 mg
6. 3.01×10^5 mg
7. 410.5 mg
8. 200 000 mg
9. 13 520 mg
10. 0.46 mg
11. 340 mg
12. 0.000 52 mg

Group II.

1. 93.8 cg
2. 1 930 cg
3. 22.8 cg
4. 1.43×10^6 cg
5. 24 320 cg
6. 2.8 cg
7. 85 cg
8. 3.1×10^4 cg
9. 6.3×10^3 cg
10. 810 cg
11. 8.012×10^{10} cg
12. 42 000 cg

Group III.

1. 7.345 6 dg
2. 0.369 dg
3. 0.41 dg
4. 15.2 dg
5. 1.24 dg
6. 1 428.1 dg
7. 8 950 dg
8. 2.3 dg
9. 730 dg
10. 0.002 38 dg

Group IV.

1. 0.415 g
2. 0.003 89 g
3. 978.5 g
4. 9 900 g
5. 119 g
6. 0.285 g
7. 0.753 g
8. 0.0145 87 g
9. 2.83 g
10. 23 g
11. 0.423 g
12. 0.013 4 g

Group V.

1. 2.450 kg
2. 14.72 kg
3. 0.089 9 kg
4. 2.38×10^{-8} kg
5. 5×10^{-7} kg
6. 2.38×10^{-4} kg
7. 0.654 kg
8. 2.4×10^2 kg
9. 1.47 kg
10. 0.6 kg

Group VI.

1. 0.014 kg
2. 0.135 4 kg
3. 7.19×10^{-5} kg
4. 51 kg
5. 5.83 kg
6. 1.42×10^{-5} kg
7. 6.2×10^{-4} kg
8. 5.2×10^{-5} kg
9. 0.039 kg
10. 4.2×10^{-6} kg

Group VII.

1. 237 g
2. 0.008 7 g
3. 0.329 g
4. 1 070 g

5. 0.039 82 g
6. 1.2×10^{-6} g
7. 4 300 g
8. 0.465 g

9. 30 g
10. 0.743 g

Group VIII.

1. 0.182 dm^3
2. 3.4×10^9 dm^3
3. 4.2×10^{12} dm^3
4. 0.000 895 dm^3

5. 0.135 4 dm^3
6. 7.5×10^{11} dm^3
7. 0.022 9 dm^3
8. 4.32×10^{-5} dm^3

9. 1.58×10^{11} dm^3
10. 2.8×10^7 dm^3
11. 6.0×10^7 dm^3
12. 0.250 dm^3

Group IX.

1. 14.2 mL
2. 0.002 590 1 mL
3. 9 230 mL
4. 7 200 mL

5. 0.323 mL
6. 1.48×10^{11} mL
7. 0.000 416 mL
8. 4.38×10^{12} mL

9. 243 000 mL
10. 0.001 079 mL

Group X.

1. 1.430×10^{-5} dm^3
2. 0.243 dm^3
3. 70 500 dm^3
4. 0.055 5 dm^3

5. 3.0×10^{-6} dm^3
6. 386 900 dm^3
7. 6.03×10^{13} dm^3
8. 0.000 128 dm^3

9. 0.009 dm^3
10. 1.12×10^7 dm^3

Group XI.

1. 0.016 mm
2. 854 mm
3. 4.32×10^{-5} mm
4. 5.8×10^{-5} mm

5. 7 200 mm
6. 0.001 25 mm
7. 2.345×10^8 mm
8. 1 420 mm

9. 319 mm
10. 7 680 mm
11. 0.006 mm
12. 304 mm

Group XII.

1. 36.7 cm
2. 12.4 cm
3. 0.000 94 cm
4. 6 000 cm

5. 2.0×10^3 cm
6. 360 cm
7. 0.001 1 cm
8. 8 590 cm

9. 85 cm
10. 1.13×10^6 cm
11. 7.41×10^{-3} cm
12. 1.047×10^{-5} cm

Group XIII.

1. 145 dm
2. 0.051 5 dm
3. 570 dm
4. 1.44 dm

5. 3.21×10^6 dm
6. 1.4×10^{-5} dm
7. 0.001 06 dm
8. 6.7×10^{-5} dm

9. 6.28×10^{-4} dm
10. 420 dm

Group XIV.

1. 1 000 m
2. 1.0×10^3 m
3. 0.785 m
4. 3.90 m
5. 1×10^{-5} m
6. 0.003 44 m
7. 468 m
8. 0.07 m
9. 6.808×10^{-7} m
10. 99 m

Group XV.

1. $2.389\ 6 \times 10^{-4}$ km
2. 5.62×10^{-8} km
3. 4.12 km
4. 0.181 km
5. 0.670 km
6. 3.71×10^5 km
7. 0.074 km
8. 0.007 63 km
9. 2.3×10^{-7} km
10. 0.043 5 km

Group XVI.

1. 24.5 cm
2. 43.8 cm
3. 0.004 15 cm
4. 7.10 cm
5. 64 cm
6. 187.44 cm
7. 830 cm
8. 0.037 9 cm
9. 311 cm
10. 2.438×10^{-5} cm

Group XVII.

1. 56.7 dm
2. 0.280 dm
3. 70 850 dm
4. 4.03 dm
5. 50.7 dm
6. 0.04 dm
7. 31 430 dm
8. 6 210 dm
9. 0.673 dm
10. 12.1 dm

Group XVIII.

1. 0.013 km
2. 0.006 5 km
3. 2.7 km
4. 3.06×10^{-5} km
5. 0.028 9 km
6. 0.218 km
7. 4.133×10^{-10} km
8. 74.5 km
9. 5.1×10^{-7} km
10. 0.160 4 km

Group XIX.

1. 1 000 μm
2. 3.29×10^6 μm
3. 5 800 μm
4. 35 000 μm
5. 1.107 5 μm
6. 4 000 μm
7. 0.481 2 μm
8. 6.9×10^{-9} μm
9. 4.28×10^{-2} μm
10. 1.29×10^5 μm

Group XX.

1. 6.34×10^8 Å
2. 640 Å
3. 4.017×10^8 Å
4. 1.080×10^7 Å
5. 7.5×10^8 Å
6. 3.0×10^{12} Å
7. 1×10^{13} Å
8. 9.0×10^4 Å
9. 3.82×10^{10} Å
10. 1.111×10^6 Å

Group XXI.

1. 4.5×10^{12} pg	5. 2.9×10^{8} pg	9. 3.6 pg
2. 8.2×10^{6} ng	6. 1.5×10^{3} ng	10. 4.28×10^{2} ng
3. 6.5×10^{7} pg	7. 1.23×10^{5} pg	11. 2.86×10^{4} pg
4. 3.17×10^{6} ng	8. 8.85 ng	12. 0.153 ng

Group XXII.

1. 340 g	6. 0.400 0 km	11. 1 022 cm^3
2. 4.2 cm	7. 44 dm^3	12. 22 cm
3. 6.300 cm	8. 1.3×10^{-6} cm^3	13. 640 g
4. 92 cm^3	9. 160 s	14. 2.0 min
5. 600 cm^3	10. 0.071 cm	15. 695 cm^3

Energy (pp. 18–22)

Group I.

1. –241 °C	5. –230.8 °C	9. 57.2 °C
2. –215 °C	6. –265.3 °C	10. 133.1 °C
3. –267.4 °C	7. 0 °C	11. –138.5 °C
4. –133.6 °C	8. 40 °C	12. –244 °C

Group II.

1. 373 K	5. 233 K	9. 558.3 K
2. 273 K	6. 117.9 K	10. 469.6 K
3. 299 K	7. 581 K	11. 689.7 K
4. 448.6 K	8. 634 K	12. 362.5 K

Group III.

1. 283 K	5. 432.5 K	9. 336.9 K
2. 351 K	6. 208.7 K	10. 487.7 K
3. 308.6 K	7. 290 K	11. 234.5 K
4. 539.3 K	8. 379.6 K	12. 11.0 K

Group IV.

1. 3 350 J	8. 16 570 J	15. 1.251×10^{6} J
2. 4 710 J	9. 927 J	16. 20 J
3. 4 440 J	10. 2.100×10^{4} J	17. 3.342×10^{4} J
4. 4.210×10^{4} J	11. 73 000 J	18. 8.063×10^{4} J
5. 33 390 J	12. 1 180 J	19. 3.03×10^{4} J
6. 90 300 J	13. 38 190 J	20. 1.13×10^{5} J
7. 548 J	14. 50 900 J	

Group V.

1. 1 390 kJ/mol
2. Reactions b and c
3. 42 100 J
4. 93.25 J/K
5. 65.7 kJ
6. –282 kJ/mol (exothermic)
7. 285 800 J
8. –282.8 kJ
9. 1 430 kJ
10. (a) 22.1J/mol• K
    (b) 109 J/kmol
11. 28.6 kJ
12. 91.4 kJ

Group VI.

1. 3.95 kJ endothermic
2. 79.5 kJ
3. 256 kJ exothermic
4. 155 kJ
5. 88.0 kJ
6. 17.4 kJ
7. 78.9 kJ
8. 3.70 kJ
9. 142 kJ
10. 2640 kJ

Atomic Structure (pp. 23–28)

Group I.

In the following answers, the order given is number of protons, number of neutrons, number of electrons, arrangement of electrons.

1. 8, 8, 8, 2-6
2. 3, 4, 3, 2-1
3. 20, 20, 20, 2-8-8-2
4. 16, 16, 16, 2-8-6
5. 18, 22, 18, 2-8-8
6. 6, 6, 6, 2-4
7. 19, 20, 19, 2-8-8-1
8. 2, 2, 2, 2
9. 13, 14, 13, 2-8-3
10. 1, 0, 1, 1

11. 4, 5, 4, 2-2
12. 9, 10, 9, 2-7
13. 7, 7, 7, 2-5
14. 33, 42, 33, 2-8-18-5
15. 83, 126, 83, 2-8-18-32-18-5
16. 79, 118, 79, 2-8-18-32-18-1
17. 44, 57, 44, 2-8-18-15-1
18. 78, 117, 78, 2-8-18-32-17-1
19. 27, 32, 27, 2-8-15-2
20. 21, 24, 21, 2-8-9-2

Group II.

Refer to Appendix D.

Group III.

In the following answers, the quantum numbers are given in the order n, l, m, s. Where a choice of possible answers is available, the answers are placed within parentheses.

1. $1, 0, 0, \pm \frac{1}{2}; 2, 0, 0, \pm \frac{1}{2}; 2, 1, 0 + \frac{1}{2}$ or $- \frac{1}{2}$
2. $1, 0, 0, \pm \frac{1}{2}; 2, 0, 0, \pm \frac{1}{2}; 2, 1, 0, \pm \frac{1}{2};$ 3 of the following $(2, 1, \pm 1 \pm \frac{1}{2})$
3. $1, 0, 0, \pm \frac{1}{2}; 2, 0, 0, \pm \frac{1}{2}; 2, 1, 0, \pm \frac{1}{2}; 2, 1, \pm 1, \pm \frac{1}{2}; 3, 0, 0, + \frac{1}{2}$ or $- \frac{1}{2}$
4. $1, 0, 0, \pm \frac{1}{2}; 2, 0, 0, \pm \frac{1}{2}; 2, 1, 0, \pm \frac{1}{2}; 2, 1, \pm 1, \pm \frac{1}{2}; 3, 0, 0, \pm \frac{1}{2}; 3, 1, 0, \pm \frac{1}{2}; 3, 1, \pm 1, \pm \frac{1}{2}$
5. $1, 0, 0, \pm \frac{1}{2}; 2, 0, 0, \pm \frac{1}{2}; 2, 1, 0, \pm \frac{1}{2};$ 2 of the following $(2, 1, \pm 1, \pm \frac{1}{2})$
6. $1, 0, 0, \pm \frac{1}{2}; 2, 0, 0, \pm \frac{1}{2}$
7. $1, 0, 0, \pm \frac{1}{2}; 2, 0, 0, \pm \frac{1}{2}; 2, 1, 0, \pm \frac{1}{2}; 2, 1, \pm 1, \pm \frac{1}{2}; 3, 0, 0, \pm \frac{1}{2}; 3, 1, 0, \pm \frac{1}{2}; 3, 1, \pm 1, \pm \frac{1}{2}; 4, 0, 0, \pm \frac{1}{2}$ or $- \frac{1}{2}$
8. $1, 0, 0, \pm \frac{1}{2}$ or $- \frac{1}{2}$
9. $1, 0, 0, \pm \frac{1}{2}; 2, 0, 0, \pm \frac{1}{2}; 2, 1, 0, \pm \frac{1}{2}; 2, 1, \pm 1, \pm \frac{1}{2}; 3, 0, 0, \pm \frac{1}{2}$
10. $1, 0, 0, \pm \frac{1}{2}; 2, 0, 0, \pm \frac{1}{2}$
11. $1, 0, 0, \pm \frac{1}{2}; 2, 0, 0, + \frac{1}{2}$ or $- \frac{1}{2}$
12. $1, 0, 0, \pm \frac{1}{2}; 2, 0, 0, \pm \frac{1}{2}; 2, 1, 0, \pm \frac{1}{2}$

Group IV.

In the following answers, p stands for protons and n for neutrons.

1. 1 p, 0 n; 1 p, 1 n
2. 3 p, 3 n; 3 p 4 n
3. 5 p, 5 n; 5 p, 6 n
4. 17 p, 18 n; 17 p, 20 n
5. 18 p, 17 n; 18 p, 20 n; 18 p, 22 n

(continued)

6. 32 p, 38 n; 32 p, 40 n; 32 p, 41 n; 32 p, 42 n; 32 p, 44 n
7. 92 p, 142 n; 92 p, 146 n
8. 52 p, 70 n; 52 p, 72 n; 52 p, 74 n; 52 p, 76 n
9. 54 p, 75 n; 54 p, 76 n; 54 p, 77 n; 54 p, 78 n; 54 p, 80 n; 54 p, 82 n
10. 63 p, 88 n; 63 p, 90 n

Group V.

Note: Isotopic compositions have been altered slightly so as to give atomic masses somewhat different from those found elsewhere.

1. 1.000 15
2. 6.925 8
3. 10.802 2
4. 35.489 4
5. 39.981 9
6. 72.715 6
7. 237.972
8. 127.785
9. 131.475
10. 151.522

Group VI.

1.

(a) 8   (c) 1   (e) 2   (g) 3
(b) 5   (d) 7   (f) 4   (h) 6

2.

(a) [Xe] $4f^{14}5d^{10}6s^26p^3$ (d) [Xe] $6s^1$ (g) [Ar] $3d^54s^2$
(b) [Xe] $4f^{14}5d^{10}6s^2$ (e) [Kr] $5s^1$ (h) [Ar]
(c) [Xe]   (f) [Kr]

3.

(a) N   (c) P   (e) K   (g) Pm
(b) Be   (d) Mn   (f) Zr   (h) Se

4.

(a) [Ar] $3d^{10}4s^1$; [Ar] $3d^{10}$; [Ar] $3d^9$
(b) [Ne] $3s^23p^1$; [Ne]
(c) [Ne] $3s^23p^5$; [Ne] $3s^23p^6$
(d) [He] $2s^22p^4$; [He] $2s^22p^6$; [He] $2s^22p^5$
(e) [Ne] $3s^23p^4$; [Ne] $3s^23p^2$; [Ne] $3s^2$; [Ne]; [Ne] $3s^23p^6$
(f) [Ne] $3s^23p^3$; [Ne]; [Ne] $3s^2$; [Ne] $3s^23p^6$

5.

	1s	2s	2p	3s	3p	4s
(a) potassium:	↑↓	↑↓	↑↓ ↑↓ ↑↓	↑↓	↑↓ ↑↓ ↑↓	↑

(b) helium:

1s
↑↓

(c) beryllium ion:

1s
[↑↓]

(d) nitrogen:

1s	2s	2p
[↑↓]	[↑↓]	[↑] [↑] [↑]

(e) fluoride ion:

1s	2s	2p
[↑↓]	[↑↓]	[↑↓] [↑↓] [↑]

(f) sulfur:

1s	2s	2p	3s	3p
[↑↓]	[↑↓]	[↑↓] [↑↓] [↑↓]	[↑↓]	[↑↓] [↑] [↑]

(g) silicon:

1s	2s	2p	3s	3p
[↑↓]	[↑↓]	[↑↓] [↑↓] [↑↓]	[↑↓]	[↑] [↑] [ ]

(h) neon:

1s	2s	2p
[↑↓]	[↑↓]	[↑↓] [↑↓] [↑↓]

6.

(a) $\cdot \overset{\displaystyle \cdot}{\underset{\displaystyle \cdot}{C}} \cdot$

(b) $\cdot \underset{\displaystyle \cdot}{Al} \cdot$

(c) $\vdots \overset{\displaystyle \cdot}{B} \cdot$

(d) $H \cdot$

(e) $\vdots \overset{\displaystyle \cdot \cdot}{\underset{\displaystyle \cdot \cdot}{F}} \cdot$

(f) $\vdots \overset{\displaystyle \cdot \cdot}{\underset{\displaystyle \cdot \cdot}{Ne}} \vdots$

(g) $Na^{\displaystyle \bullet}$

(h) K^+

(i) $\vdots \overset{\displaystyle \cdot \cdot}{\underset{\displaystyle \cdot \cdot}{S}} \vdots \ ^{2-}$

(j) Mg^{2+}

(k) B^{3+}

(l) $\vdots \overset{\displaystyle \cdot \cdot}{\underset{\displaystyle \cdot \cdot}{Ar}} \vdots$

7.

(a) oxygen:

2s	2p
[↑↓]	[↑↓] [↑] [↑]

(b) boron:

2s	2p
[↑↓]	[↑] [↑] [ ]

(c) aluminum:

3s	3p
[↑↓]	[↑] [ ] [ ]

(d) calcium:

4s
[↑↓]

(e) Na^+:

2s	2p
[↑↓]	[↑↓] [↑↓] [↑↓]

(f) Br^-:

4s	4p
[↑↓]	[↑↓] [↑↓] [↑↓]

(g) Mg^{2+}:

2s	2p
[↑↓]	[↑↓] [↑↓] [↑↓]

(h) S^{2-}:

3s	3p
[↑↓]	[↑↓] [↑↓] [↑↓]

Bonding Reactions (pp. 29–30)

Note: Electron dot formulas below may not represent bond types actually formed in some cases.

1. Na^* + $.\ddot{C}\ddot{l}:$ → Na^+ + $^*_.\ddot{C}\ddot{l}:^-$ $\Delta E_n = 1.9$

2. K^* + $.\ddot{F}:$ → K^+ + $^*_.\ddot{F}:^-$ $\Delta E_n = 3.2$

3. Li^* + $.\ddot{S}:$ → $Li^*_.\ddot{S}:$ $\Delta E_n = 1.4$
   Li^* Li

4. $4\,H^*$ + $.\dot{C}.$ → $H^*_.C^*_.H$ (with H above and below) $\Delta E_n = 0.4$

5. Be^*_* + $\ddot{S}:$ → $Be^*_*\ddot{S}:$ $\Delta E_n = 0.9$

6. Mg^*_* +   → Mg^{2+} + $\Delta E_n = 1.8$

   Mg^*_* $\ddot{N}:$ Mg^{2+} $^*_*N:^{3-}$
   Mg^*_* $\ddot{N}:$ Mg^{2+} $^*_*N:^{3-}$

7. $4\,.\ddot{F}:$ + $.\dot{S}i.$ → Si^{4+} + $4\ ^*_*\ddot{F}:^-$ $\Delta E_n = 2.3$

8. Al^* (with ** above) + $.\dot{N}.$ → $Al^*_*\ddot{N}^*_*$ $\Delta E_n = 1.6$

(continued)

9. 4 :Cl: +   C   →   Cl C Cl (with Cl above and below) $\Delta E_n = 0.4$

10. 4 :F: +   C   →   F C F (with F above and below) $\Delta E_n = 1.6$

11. S   +   3 :O: →   O S (with O below, two O) $\Delta E_n = 1.1$

12. 4 H   +   Ge   →   H Ge H (with H above and below) $\Delta E_n = 0.1$

13. Ca   +   :Cl: (two)   → Ca^{2+} +   :Cl:$^{1-}$ (two) $\Delta E_n = 1.8$

14. Mg   Mg   +   P   P   →   Mg P :: P Mg (with Mg above) $\Delta E_n = 0.8$

Mg   Mg   P

15. Zn   +   :Br: (two)   →   Br Zn Br $\Delta E_n = 1.1$

Formulas and Nomenclature (pp. 31–36)

Group I.

1. hydrogen chloride or hydrochloric acid
2. potassium hydroxide
3. mercury(I) hydroxide
4. potassium chloride
5. iron(III) chloride
6. nitric acid or hydrogen nitrate
7. ammonium hydroxide
8. copper(I) oxide
9. aluminum sulfate
10. dinitrogen pentoxide
11. sodium hydroxide
12. carbon dioxide
13. hydrofluoric acid or hydrogen fluoride
14. lead(II) hydroxide
15. ammonium nitrate
16. sodium bicarbonate or sodium hydrogen carbonate
17. mercury(II) oxide
18. zinc nitrite
19. phosphoric acid or hydrogen phosphate
20. cesium hydroxide
21. lithium oxide
22. calcium hydroxide
23. calcium bromide
24. iron(III) oxide
25. sulfuric acid or hydrogen sulfate
26. iron(II) carbonate
27. sulfur trioxide
28. barium bromate
29. aluminum hydroxide
30. perchloric acid or hydrogen perchlorate
31. sodium acetate
32. sodium sulfite
33. carbonic acid or hydrogen carbonate
34. fluorous acid or hydrogen fluorite
35. ammonium iodate
36. lithium hydride
37. carbon monoxide

(continued)

38. magnesium bromide
39. tin(IV) bromide
40. nitrous oxide
41. ammonium fluoride
42. arsenic pentachloride
43. potassium bicarbonate
44. potassium oxide
45. barium arsenide
46. zinc oxide
47. sodium hypochlorite
48. strontium sulfide
49. aluminum bromate
50. antimony trifluoride
51. palladium cyanide
52. zinc silicate
53. magnesium acetate
54. calcium permanganate
55. beryllium nitrate
56. nickel selenate
57. radium bromide
58. sodium permanganate
59. lead(II) iodide
60. calcium sulfide
61. bismuth telluride
62. potassium perchlorate
63. mercury(II) bromide
64. cobalt silicide
65. triphosphorus pentanitride
66. copper(II) sulfite
67. iron(III) phosphate
68. lead(II) telluride
69. mercury(I) nitrate
70. potassium silicate
71. silver acetate
72. tellurium tetraiodide
73. zinc phosphate
74. silver sulfide
75. cadmium bicarbonate
76. zinc fluoride
77. sulfurous acid or hydrogen sulfite

(continued)

78. barium hydroxide
79. lead(II) sulfide
80. sodium dihydrogen phosphate or monobasic sodium phosphate
81. ammonium acetate
82. silver nitride
83. silicon tetraiodide
84. zinc carbonate
85. phosphorus acid or hydrogen phosphite
86. tin(IV) iodide
87. lead(II) nitrate
88. sodium fluoride
89. potassium aluminum sulfate
90. potassium uranate
91. samarium chloride
92. potassium pentasulfide
93. iron(II) ferricyanide
94. platinum(II) chloride
95. platinum(IV) iodide
96. nitrogen triiodide
97. molybdenum pentachloride
98. lanthanum nitrate
99. dysprosium oxide
100. vanadium pentoxide

Group II.

1. H_2SO_4
2. $NaOH$
3. $NaBr$
4. $Ba(OH)_2$
5. CaO
6. H_2S
7. Li_2SO_4
8. CO
9. MnO_2
10. SO_2
11. $FeSO_4$
12. $HClO$
13. $KMnO_4$
14. $AgCl$
15. $Cu(OH)_2$
16. $(NH_4)_2S$
17. $NiBr_2$
18. FeO
19. $HBrO_3$
20. NH_4HSO_4
21. Hg_2SO_4
22. Fe_2O_3
23. $Mg_3(PO_4)_2$
24. $Ni(HCO_3)_2$
25. $Zn(OH)_2$
26. HI
27. P_2O_5
28. $AlPO_4$
29. $HC_2H_3O_2$
30. $Cu(NO_2)_2$
31. NO_2
32. PCl_3
33. Na_3PO_4
34. K_2CO_3
35. H_3PO_4
36. $PbCl_4$
37. $SnBr_2$
38. NH_4OH
39. HIO_4
40. $Fe(OH)_2$
41. CO_2
42. N_2O_5
43. Ag_2O
44. AlN
45. $Mn(OH)_2$
46. $(NH_4)_2CO_3$
47. Al_2O_3
48. Sb_2S_5
49. $BaCO_3$
50. $Ca_3(PO_4)_2$
51. Cs_2CO_3
52. K_2SiO_3
53. Ag_2CrO_4
54. $MgSO_3$
55. CrP
56. $Co(NO_3)_2$

(continued)

57. ZnI_2
58. FeF_2
59. $NiSe$
60. $NaHSO_4$
61. Li_2O
62. Cu_2CO_3
63. $SrCO_3$
64. Hg_2SO_4
65. $K_2Cr_2O_7$
66. MnO
67. $NiCl_2$
68. $Pb(C_2H_3O_2)_2$
69. Hg_3N_2
70. $Pb(OH)_2$

71. $SnCl_4$
72. SeF_4
73. PBr_5
74. $HgIO_3$
75. $Fe_2(SO_4)_3$
76. $NiSO_4$
77. SiO_2
78. Li_3PO_4
79. K_3Sb
80. HNO_3
81. Mg_3N_2
82. $Cd(NO_2)_2$
83. $Zn(C_2H_3O_2)_2$
84. HNO_2

85. $Sr(OH)_2$
86. $PbSO_4$
87. $Al(HSO_4)_3$
88. Na_2HPO_4
89. $NH_4Al(SO_4)_2$
90. $CuSO_4 \cdot 5H_2O$
91. $Pb(NO_3)_2$
92. $AuCl_3$
93. $Sn(OH)_2$
94. H_2CO_3
95. NH_4BrO_3
96. $ScBr_3$
97. BrI
98. Rb_2CO_3

99. $K_2S_2O_3$
100. K_3AsO_4
101. $KAg(CN)_2$
102. $NaCNO$
103. $HMnO_4$
104. $OsCl_4$
105. La_2O_3
106. $GeCl_4$
107. $Er(C_2H_3O_2)_3$
108. Yb_2O_3
109. CaH_2
110. $Fe_3[Fe(CN)_6]_2$

Equations (pp. 37–44)

1. $Fe + S \rightarrow FeS$
2. $Zn + CuSO_4 \rightarrow ZnSO_4 + Cu$
3. $AgNO_3 + NaBr \rightarrow NaNO_3 + AgBr$
4. $2KClO_3 \xrightarrow{\Delta} 2KCl + 3O_2 \uparrow$
5. $2H_2O \rightleftharpoons 2H_2 + O_2 \uparrow$
6. $2HgO \xrightarrow{\Delta} 2Hg + O_2 \uparrow$
7. $2KI + Pb(NO_3)_2 \rightarrow PbI_2 + 2KNO_3$
8. $4Al + 3O_2 \rightarrow 2Al_2O_3$
9. $MgCl_2 + 2NH_4NO_3 \rightarrow Mg(NO_3)_2 + 2NH_4Cl$
10. $FeCl_3 + 3NH_4OH \rightarrow Fe(OH)_3 + 3NH_4Cl$
11. $2Na_2O_2 + 2H_2O \rightarrow 4NaOH + O_2 \uparrow$
12. $Fe_2O_3 + 3C \rightarrow 2Fe + 3CO \uparrow$
13. $2Fe + 3H_2O \rightarrow 3H_2 \uparrow + Fe_2O_3$
14. $FeCl_3 + 3KOH \rightarrow 3KCl + Fe(OH)_3$
15. $2Al + 3H_2SO_4 \rightarrow Al_2(SO_4)_3 + 3H_2 \uparrow$
16. $Na_2CO_3 + Ca(OH_2) \rightarrow 2NaOH + CaCO_3$
17. $CO_2 + H_2O \rightarrow H_2CO_3$
18. $4P + 5O_2 \rightarrow 2P_2O_5$
19. $2Na + 2HOH \rightarrow 2NaOH + H_2 \uparrow$
20. $Zn + H_2SO_4 \rightarrow ZnSO_4 + H_2 \uparrow$
21. $Al_2(SO_4)_3 + 3Ca(OH)_2 \rightarrow 2Al(OH)_3 + 3CaSO_4$
22. $CaO + H_2O \rightarrow Ca(OH)_2$
23. $Fe + 2CuNO_3 \rightarrow Fe(NO_3)_2 + 2Cu$

(continued)

24. $FeS + 2HCl \rightarrow H_2S \uparrow + FeCl_2$

25. $K_2O + H_2O \rightarrow 2KOH$

26. $(NH_4)_2S + Pb(NO_3)_2 \rightarrow 2NH_4NO_3 + PbS$

27. $3Hg(OH)_2 + 2H_3PO_4 \rightarrow Hg_3(PO_4)_2 + 6H_2O$

28. $3KOH + H_3PO_4 \rightarrow K_3PO_4 + 3H_2O$

29. $CaCl_2 + 2\ HNO_3 \rightarrow Ca(NO_3)_2 + 2\ HCl$

30. $K_2CO_3 + BaCl_2 \rightarrow 2KCl + BaCO_3$

31. $Mg(OH)_2 + H_2SO_4 \rightarrow MgSO_4 + 2H_2O$

32. $SO_2 + H_2O \rightarrow H_2SO_3$

33. $Na_2CO_3 + 2HCl \rightarrow 2NaCl + H_2O + CO_2 \uparrow$

34. $Mg + 2HNO_3 \rightarrow Mg(NO_3)_2 + H_2 \uparrow$

35. $2Al + Fe_2O_3 \rightarrow Al_2O_3 + 2Fe$

36. $2K_3PO_4 + 3MgCl_2 \rightarrow Mg_3(PO_4)_2 + 6KCl$

37. $4NH_3 + 3O_2 \rightarrow 2N_2 \uparrow + 6H_2O$

38. $CaCO_3 \overset{\Delta}{\rightarrow} CaO + CO_2 \uparrow$

39. $2NaCl + H_2SO_4 \rightarrow Na_2SO_4 + 2HCl$

40. $2F_2 + 4NaOH \rightarrow 4NaF + O_2 \uparrow + 2H_2O$

41. $Mg(NO_3)_2 + CaI_2 \rightarrow Ca(NO_3)_2 + MgI_2$

42. $Al_2(SO_4)_3 + 6NH_4Br \rightarrow 2AlBr_3 + 3(NH_4)_2SO_4$

43. $2KF + BaBr_2 \rightarrow BaF_2 + 2KBr$

44. $Cu(NO_3)_2 + 2NH_4OH \rightarrow Cu(OH)_2 + 2NH_4NO_3$

45. $2NaNO_3 \overset{\Delta}{\rightarrow} 2NaNO_2 + O_2 \uparrow$

46. $Pb(OH)_2 \overset{\Delta}{\rightarrow} PbO + H_2O$

47. $2NH_3 + H_2SO_4 \rightarrow (NH_4)_2SO_4$

48. $HCl + NH_3 \rightarrow NH_4Cl$

49. $CuSO_4 + Fe \rightarrow FeSO_4 + Cu$

50. $2Al + 6HCl \rightarrow 2AlCl_3 + 3H_2 \uparrow$

51. $C + O_2 \rightarrow CO_2 \uparrow$

52. $Ca(HCO_3)_2 + Ca(OH)_2 \rightarrow 2CaCO_3 + 2H_2O$

53. $2H_2S + O_2 \rightarrow 2H_2O + 2S$

54. $2NaOH + Ca(NO_3)_2 \rightarrow 2NaNO_3 + Ca(OH)_2$

55. $2KI + Cl_2 \rightarrow 2KCl + I_2$

56. $H_2SO_4 + 2KOH \rightarrow K_2SO_4 + 2H_2O$

57. $CO_2 + C \rightarrow 2CO \uparrow$

58. $CaSO_4 + Na_2CO_3 \rightarrow CaCO_3 + Na_2SO_4$

59. $3H_2O + P_2O_5 \rightarrow 2H_3PO_4$

60. $2Al + 2H_3PO_4 \rightarrow 3H_2 \uparrow + 2AlPO_4$

61. $NH_4Cl + NaNO_2 \rightarrow NaCl + N_2 \uparrow + H_2O$

62. $Cl_2 + 2NaOH \rightarrow NaCl + NaClO + H_2O$

63. $2Pb(NO_3)_2 \overset{\Delta}{\rightarrow} 2PbO + 4NO_2 \uparrow + O_2 \uparrow$

(continued)

64. $2Hg_2O + O_2 \rightarrow 4HgO$

65. $CaO + MgCl_2 \rightarrow MgO + CaCl_2$

66. $Ca + 2H_2O \rightarrow Ca(OH)_2 + H_2 \uparrow$

67. $2CrCl_3 + 3H_2SO_4 \rightarrow Cr_2(SO_4)_3 + 6HCl$

68. $Fe(NO_3)_3 + 3NH_4OH \rightarrow Fe(OH)_3 + 3NH_4NO_3$

69. $AlCl_3 + K_3PO_4 \rightarrow AlPO_4 + 3KCl$

70. $Al_2O_3 + 3C + 3Cl_2 \rightarrow 3CO \uparrow + 2AlCl_3$

71. $Cu_2O + 2HCl \rightarrow 2CuCl + H_2O$

72. $Mg(HCO_3)_2 + 2HCl \rightarrow MgCl_2 + 2H_2O + 2CO_2 \uparrow$

73. $4Fe + 3O_2 \rightarrow 2Fe_2O_3$

74. $Si + 2H_2O \xrightarrow{\Delta} SiO_2 + 2H_2 \uparrow$

75. $Fe_2O_3 + 3CO \rightarrow 2Fe + 3CO_2 \uparrow$

76. $3CaCl_2 + 2Cr(NO_3)_3 \rightarrow 3Ca(NO_3)_2 + 2CrCl_3$

77. $2ZnS + 3O_2 \rightarrow 2ZnO + 2SO_2 \uparrow$

78. $Ca_3(PO_4)_2 + 3H_2SO_4 \rightarrow 3CaSO_4 + 2H_3PO_4$

79. $2Fe(OH)_3 \xrightarrow{\Delta} Fe_2O_3 + 3H_2O$

80. $Al_2S_3 + 6NaHCO_3 \rightarrow 2Al(OH)_3 + 3Na_2SO_4 + 6CO_2 \uparrow$

81. $Ca_3(PO_4)_2 + 3SiO_2 + 5C \rightarrow 2P + 3CaSiO_3 + 5CO \uparrow$

82. $CaO + SO_2 \rightarrow CaSO_3$

83. $CO_2 + Mg(OH)_2 \rightarrow MgCO_3 + H_2O$

84. $CaO + 2HCl \rightarrow CaCl_2 + H_2O$

85. $CaCO_3 + SiO_2 \rightarrow CaSiO_3 + CO_2 \uparrow$

86. $2Sb + 3Cl_2 \rightarrow 2SbCl_3$

87. $Mg_3N_2 + 6H_2O \rightarrow 3Mg(OH)_2 + 2NH_3 \uparrow$

88. $4As + 3O_2 \rightarrow 2As_2O_3$

89. $NH_4HCO_3 \xrightarrow{\Delta} NH_3 \uparrow + H_2O \uparrow + CO_2 \uparrow$

90. $3CuO + 2NH_3 \rightarrow 3Cu + 3H_2O + N_2 \uparrow$

91. $(NH_4)_2Cr_2O_7 \xrightarrow{\Delta} Cr_2O_3 + N_2 \uparrow + 4H_2O$

92. $H_2S + Cd(NO_3)_2 \rightarrow 2HNO_3 + CdS$

93. $3BaBr_2 + 2Na_3PO_4 \rightarrow Ba_3(PO_4)_2 + 6NaBr$

94. $AlCl_3 + 3NH_4F \rightarrow AlF_3 + 3NH_4Cl$

95. $2AgNO_3 + K_2SO_4 \rightarrow Ag_2SO_4 + 2KNO_3$

96. $2Bi(NO_3)_3 + 3CaI_2 \rightarrow 2BiI_3 + 3Ca(NO_3)_2$

97. $Al_2(CrO_4)_3 + 3(NH_4)_2SO_4 \rightarrow 3(NH_4)_2CrO_4 + Al_2(SO_4)_3$

98. $Zn(NO_3)_2 + 2NH_4Br \rightarrow ZnBr_2 + 2NH_4NO_3$

99. $Bi(NO_3)_3 + 3NH_4OH \rightarrow Bi(OH)_3 + 3NH_4NO_3$

100. $Cd(NO_3)_2 + H_2SO_4 \rightarrow CdSO_4 + 2HNO_3$

101. $Zn + 2AgI \rightarrow ZnI_2 + 2Ag$

102. $2FeCl_3 + 3H_2SO_4 \rightarrow Fe_2(SO_4)_3 + 6HCl$

103. $Bi_2(SO_4)_3 + 6NH_4OH \rightarrow 2Bi(OH)_3 + 3(NH_4)_2SO_4$

(continued)

104. $4HI + O_2 \rightarrow 2I_2 + 2H_2O$

105. $K_2SO_4 + BaCl_2 \rightarrow BaSO_4 + 2KCl$

106. $BaSO_4 + 4C \rightarrow BaS + 4CO \uparrow$

107. $Al_2O_3 + 6HF \rightarrow 2AlF_3 + 3H_2O$

108. $2AlF_3 + 3H_2SO_4 \rightarrow Al_2(SO_4)_3 + 6HF \uparrow$

109. $2KI + H_2O_2 \rightarrow 2KOH + I_2$

110. $Zn + Fe_2(SO_4)_3 \rightarrow ZnSO_4 + 2FeSO_4$

111. $PbS + 2PbO \rightarrow 3Pb + SO_2 \uparrow$

112. $Cu + 2H_2SO_4 \rightarrow CuSO_4 + 2H_2O + SO_2 \uparrow$

113. $2Al(OH)_3 \overset{\Delta}{\rightarrow} Al_2O_3 + 3H_2O$

114. $N_2 + 3H_2 \rightarrow 2NH_3 \uparrow$

115. $Na_2CO_3 + H_2CO_3 \rightarrow 2NaHCO_3$

116. $SiO_2 + 4HF \rightarrow 2H_2O + SiF_4 \uparrow$

117. $3NaClO \rightarrow 2NaCl + NaClO_3$

118. $2NaClO_2 + Cl_2 \rightarrow 2NaCl + 2ClO_2 \uparrow$

119. $CH_4 + SO_2 \overset{\Delta}{\rightarrow} H_2S \uparrow + CO_2 \uparrow + H_2 \uparrow$

120. $H_2TeO_3 \rightarrow TeO_2 \uparrow + H_2O$

121. $FeSe + 2HCl \rightarrow FeCl_2 + H_2Se \uparrow$

122. $3Mg + N_2 \rightarrow Mg_3N_2$

123. $AgCN + K \rightarrow KCN + Ag$

124. $CuSO_4 + 4NH_3 \rightarrow Cu(NH_3)_4SO_4$

125. $CaC_2 + N_2 \rightarrow CaCN_2 + C$

126. $CaCN_2 + 3H_2O \rightarrow CaCO_3 + 2NH_3 \uparrow$

127. $Zn_3As_2 + 6HCl \rightarrow 2AsH_3 \uparrow + 3ZnCl_2$

128. $Pb(OH)_2 + Na_2SnO_2 \rightarrow Pb + Na_2SnO_3 + H_2O$

129. $Na_2SiO_3 + 2HCl \rightarrow 2NaCl + H_2SiO_3$

130. $B_2O_3 + 3Mg \rightarrow 3MgO + 2B$

131. $Fe(CN)_2 + 4KCN \rightarrow K_4Fe(CN)_6$

132. $NaAlO_2 + NH_4Cl + NH_4AlO_2 + NaCl$

133. $Al(OH)_3 + 3NaOH \rightarrow Na_3AlO_3 + 3H_2O$

134. $W + 3Cl_2 \overset{\Delta}{\rightarrow} WCl_6$

135. $3Ca + 2NH_3 \rightarrow 3CaH_2 + N_2 \uparrow$

136. $Na_2WO_4 + H_2SO_4 \rightarrow Na_2SO_4 + H_2WO_4$

137. $LiH + H_2O \rightarrow LiOH + H_2 \uparrow$

138. $4H_3BO_3 \rightarrow H_2B_4O_7 + 5H_2O$

139. $Zn(OH)_2 + 2KOH \rightarrow K_2ZnO_2 + 2H_2O$

140. $Ni + 4CO \rightarrow Ni(CO)_4$

Reaction Prediction (pp. 45–50)

In the following problems, the answers are given in this order:

(a) Type of reaction: S = synthesis; D = decomposition; SR = single replacement; DR = double replacement

(b) Yes or no, as to whether the reaction occurs

(c) Balanced equation for those that do

For double replacement reactions, any *product* listed in Appendix L as P, I, A, or a is regarded as insoluble; any *reactant* listed as P is regarded as soluble.

1. SR; yes; $2Al + 6HCl \rightarrow 2AlCl_3 + 3H_2$
2. DR; yes; $Ca(OH)_2 + 2HNO_3 \rightarrow Ca(NO_3)_2 + 2H_2O$
3. S; no
4. SR; yes; $Mg + Zn(NO_3)_2 \rightarrow Mg(NO_3)_2 + Zn \downarrow$
5. S; yes; $2Hg + O_2 \rightarrow 2HgO$
6. DR; yes; $ZnCl_2 + H_2S \rightarrow 2HCl + ZnS \downarrow$
7. S; yes; $N_2O_5 + H_2O \rightarrow 2NHO_3$
8. DR; no; $AgCl \downarrow$
9. D; yes; $2NaClO_3 \overset{\Delta}{\rightarrow} 2NaCl + 3O_2 \uparrow$
10. DR; yes; $Ba(NO_3)_2 + Na_2CrO_4 \rightarrow BaCrO_4 \downarrow + 2NaNO_3$
11. DR; yes; $NaBr + AgNO_3 \rightarrow NaNO_3 + AgBr \downarrow$
12. DR; yes; $Ca_3(PO_4)_2 + Al_2(SO_4)_3 \rightarrow 3CaSO_4 + 2AlPO_4 \downarrow$
13. D; yes; $ZnCO_3 \overset{\Delta}{\rightarrow} ZnO + CO_2 \uparrow$
14. DR; yes; $Hg_2SO_4 + 2NH_4NO_3 \rightleftharpoons (NH_4)_2SO_4 + 2HgNO_3$
15. S; yes; $2K + F_2 \rightarrow 2KF$
16. DR; no; $Zn_3(PO_4)_2 \downarrow$
17. S; yes; $Li_2O + H_2O \rightarrow 2LiOH$
18. D; yes; $2NaCl \overset{\sim}{\rightsquigarrow} 2Na + Cl_2 \uparrow$
19. S; no
20. DR; yes; $Fe(OH)_3 + H_3PO_4 \rightarrow FePO_4 \downarrow + 3H_2O$
21. SR; yes; $2Na + 2HNO_3 \rightarrow 2NaNO_3 + H_2 \uparrow$
22. DR; yes; $2FeI_3 + 3Cu(NO_3)_2 \rightarrow 2Fe(NO_3)_3 + 3CuI_2$ (decomposes in water)
23. SR; no
24. SR; no
25. S; yes; $SO_2 + H_2O \rightleftharpoons H_2SO_3$
26. S; yes; $S + O_2 \rightarrow SO_2 \uparrow$
27. DR; yes; $Na_2SO_4 + BaCl_2 \rightarrow BaSO_4 \downarrow + 2NaCl$
28. DR; yes; $(NH_4)_3PO_4 + 3LiOH \rightarrow Li_3PO_4 + 3NH_4OH$ $(\rightarrow 3NH_3 \uparrow + 3H_2O)$
29. S; yes; $2H_2 + O_2 \rightarrow 2H_2O$
30. SR; no
31. S; yes; $Na_2O + H_2O \rightarrow 2NaOH$
32. DR; yes; $CaCO_3 + 2LiCl \rightleftharpoons Li_2CO_3 + CaCl_2$

(continued)

33. DR; yes; $Hg_2SO_4 + 2HCl \rightleftharpoons H_2SO_4 + 2HgCl$

34. D; yes; $2KNO_3 \xrightarrow{\Delta} 2KNO_2 + O_2 \uparrow$

35. S; may occur under certain conditions; $Cl_2 + Br_2 \rightarrow 2BrCl$

36. DR; yes; $2HgNO_3 + Na_2CO_3 \rightarrow 2NaNO_3 + Hg_2CO_3 \downarrow$

37. SR; yes; $Mg + 2HCl \rightarrow MgCl_2 + H_2 \uparrow$

38. D; yes; $2H_2O \rightleftharpoons 2H_2 \uparrow + O_2 \uparrow$

39. DR; yes; $2NH_4NO_2 + Ba(OH)_2 \rightarrow 2NH_4OH(\rightarrow 2NH_3 \uparrow + 2H_2O) + Ba(NO_2)_2$

40. DR; yes; $(NH_4)_2SO_4 + Ca(OH)_2 \rightarrow 2NH_4OH (2NH_3 \uparrow + 2H_2O) + CaSO_4$

41. D; yes; $2HgO \xrightarrow{\Delta} 2Hg + O_2 \uparrow$

42. DR; yes; $(NH_4)_3PO_4 + AlCl_3 \rightarrow AlPO_4 \downarrow + 3NH_4Cl$

43. S; yes; $BaO + H_2O \rightarrow Ba(OH)_2$

44. DR; yes; $Fe(OH)_3 + 3HNO_3 \rightarrow Fe(NO_3)_3 + 3H_2O$

45. S; yes; $2Ca + O_2 \rightarrow 2CaO$

46. SR; yes; $3Ca + 2H_3PO_4 \rightarrow Ca_3(PO_4)_2 + 3H_2 \uparrow$

47. DR; yes; $CaCl_2 + 2NH_4OH \rightleftharpoons Ca(OH)_2 + 2NH_4Cl$

48. DR; yes; provided the Al_2S_3 is added as a solid; Al_2S_3 does not exist in water solution; $Al_2S_3 + 6HCl \rightarrow 2AlCl_3 + 3H_2S \uparrow$

49. S; yes; $Mg + S \rightarrow MgS$

50. SR; yes; $3Ca + 2AlCl_3 \rightarrow 3CaCl_2 + 2Al$

51. DR; yes; $2KOH + H_2S \rightarrow 2H_2O + K_2S$

52. DR; yes; $Na_2CO_3 + H_2SO_4 \rightarrow Na_2SO_4 + H_2CO_3(H_2O + CO_2 \uparrow)$

53. DR; no; $BaSO_4 \downarrow$

54. SR; no

55. D; yes; $BaCO_3 \xrightarrow{\Delta} BaO + CO_2 \uparrow$

56. S; yes; $2Li + Br_2 \rightarrow 2LiBr$

57. DR; yes; $2NaCl + K_2CrO_4 \rightleftharpoons Na_2CrO_4 + 2KCl$

58. DR; yes; $K_2S + Fe(NO_3)_2 \rightarrow FeS \downarrow + 2KNO_3$

59. SR; no

60. S; no

61. D; yes; $2AlCl_3 \rightleftharpoons 2Al + 3Cl_2 \uparrow$

62. DR; yes; $Pb(ClO_3)_2 + Na_2S \rightarrow PbS \downarrow + 2NaClO_3$

63. S; yes, although there are practical difficulties; $SO_3 + H_2O \rightarrow H_2SO_4$

64. DR; yes; $CaCO_3 + 2HCl \rightarrow CaCl_2 + H_2CO_3 (\rightarrow H_2O + CO_2 \uparrow)$

65. SR; no

66. DR; yes; $2NH_4C_2H_3O_2 + FeCl_2 \rightleftharpoons 2NH_4Cl + Fe(C_2H_3O_2)_2$

67. DR; no; $AgBr \downarrow$

68. SR; yes; $Zn + H_2SO_4 \rightarrow ZnSO_4 + H_2 \uparrow$

69. S; no

70. SR; no

71. DR; yes; $Pb(OH)_2 + 2HCl \rightarrow PbCl_2 + 2H_2O$

(continued)

72. S; yes; Fe + S $\rightarrow$ FeS

73. D; yes; 2KClO$_3$ $\overset{\Delta}{\rightarrow}$ 3O$_2$ $\uparrow$ + 2KCl

74. S; under special circumstances, yes; Cl$_2$ + 2O$_2$ $\rightarrow$ 2ClO$_2$

75. DR; no; AgI $\downarrow$

76. DR; yes; 3FeCO$_3$ + 2H$_3$PO$_4$ $\rightarrow$ Fe$_3$(PO$_4$)$_2$ $\downarrow$ + 3H$_2$CO$_3$ ($\rightarrow$ 3H$_2$O + 3CO$_2$ $\uparrow$)

77. DR; yes; KI + NH$_4$NO$_3$ $\rightleftharpoons$ KNO$_3$ + NH$_4$I

78. SR; yes; K + NaNO$_3$ $\overset{\Delta}{\rightarrow}$ Na + KNO$_3$

79. SR; no

80. DR; yes; Ag$_2$S + 2HCl $\rightarrow$ 2AgCl $\downarrow$ + H$_2$S $\uparrow$

81. DR; yes; Mg(NO$_3$)$_2$ + 2HCl $\rightleftharpoons$ 2NHO$_3$ + MgCl$_2$

82. S; yes; NH$_3$ + HCl $\rightarrow$ NH$_4$Cl

83. DR; yes; Zn(OH)$_2$ + H$_2$SO$_4$ $\rightarrow$ ZnSO$_4$ + 2H$_2$O

84. S; yes; CaO + H$_2$O $\rightarrow$ Ca(OH)$_2$

85. S; yes; 2Na + Cl$_2$ $\rightarrow$ 2NaCl

86. D; yes; Ca(OH)$_2$ $\overset{\Delta}{\rightarrow}$ CaO + H$_2$O

87. SR; yes; F$_2$ + 2KBr $\rightarrow$ 2KF + Br$_2$

88. DR; yes; 2NH$_4$OH + H$_2$SO$_4$ $\rightarrow$ (NH$_4$)$_2$SO$_4$ + 2H$_2$O

89. DR; yes; NaCl + KNO$_3$ $\rightleftharpoons$ NaNO$_3$ + KCl

90. SR; no

91. S; yes; CO$_2$ + H$_2$O $\rightleftharpoons$ H$_2$CO$_3$

92. SR; yes; Cl$_2$ + 2LiBr $\rightarrow$ 2LiCl + Br$_2$

93. DR; yes; 3LiOH + H$_3$PO$_4$ $\rightarrow$ Li$_3$PO$_4$ + 3H$_2$O

94. DR; yes; K$_2$SO$_3$ + 2HNO$_3$ $\rightarrow$ 2KNO$_3$ + H$_2$SO$_3$ ($\rightarrow$ H$_2$O + SO$_2$ $\uparrow$)

95. DR; yes; NH$_4$Cl + KOH $\rightarrow$ KCl + NH$_4$OH($\rightarrow$ NH$_3$ $\uparrow$ + H$_2$O)

96. DR; yes; SrCO$_3$ + 2HNO$_3$ $\rightarrow$ Sr(NO$_3$)$_2$ + H$_2$CO$_3$ ($\rightarrow$ H$_2$O + CO$_2$ $\uparrow$)

97. SR; yes; Sn + 2HgNO$_3$ $\rightarrow$ Sn(NO$_3$)$_2$ + 2Hg

98. DR; yes; (NH$_4$)$_2$SO$_3$ + 2HCl $\rightarrow$ 2NH$_4$Cl + H$_2$SO$_3$ ($\rightarrow$ H$_2$O + SO$_2$ $\uparrow$)

99. DR; yes; 3MgCO$_3$ + 2H$_3$PO$_4$ $\rightarrow$ Mg$_3$(PO$_4$)$_2$ + 3H$_2$CO$_3$ ($\rightarrow$ 3H$_2$O + 3CO$_2$ $\uparrow$)

100. DR; yes; Al$_2$(SO$_3$)$_3$ + 6HCl $\rightarrow$ 2AlCl$_3$ + 3H$_2$SO$_3$ ($\rightarrow$ 3H$_2$O + 3CO$_2$ $\uparrow$)

Density and Specific Gravity (pp. 51–54)

- SOLIDS AND LIQUIDS

1. 8.9
2. 21.4 g/cm^3, 21.4
3. 11.3 g/cm^3, 11.3
4. 1500 g
5. 0.900
6. 2.7, 2.7 g/cm^3
7. 1200 g
8. 2.70
9. 130 cm^3
10. 480 g
11. 1.3
12. 8500 kg
13. 1.84
14. 0.991 g/cm^3
15. 7 100 kg/m^3
16. 100 g (99.7 g)
17. 1 260 g
18. 0.920 g/cm^3, 0.920
19. 150 mL
20. 68 000 g

- GASES AND VAPORS

1. 1.639 g/L, 1.268
2. 0.7710 g/L, 0.5963
3. 44.303, 1.5297
4. 1.25 g/L, 28.0
5. 3.21 g/L, 72.1
6. 0.0695, 2.01
7. 3.74 g/L, 2.89, 83.7
8. 10. g
9. 5.61 g/L, 126
10. 4.61 g/L, 3.57, 103
11. 3.165 g/L
12. 3.02, 87.4
13. 1.206 g
14. .554
15. 5.70 g/L, 4.41, 128
16. 5.517
17. 0.586 g
18. 2.72 g/L, 60.8
19. 1.62, 46.8
20. 2.927 g/L, 2.264, 65.56
21. 0.1381, 4.00
22. 0.6830 g
23. 1.341 g/L, 30.03
24. 1.38
25. 1.43 g/L, 1.11, 32.0
26. 1.65 g
27. 1.25 g/L, 0.967
28. 28.6
29. 23.0 g
30. 3.61 g/L, 2.79

Gas Law Problems (pp. 55–62)

Group 1.

1. 140 mL
2. 330 cm^3
3. 290 mL
4. 0.984 L
5. 97.33 cm^3
6. 244 mL
7. 38.0 cm^3
8. 656 mL
9. 431 cm^3
10. 643 mL
11. 73.5 cm^3
12. 40 mL
13. 58.6 cm^3
14. 32.5 mL
15. 11.4 cm^3
16. 252 mL
17. 244 cm^3
18. 45 mL
19. 136 cm^3
20. 262 mL
21. 1.7 dm^3
22. 683 cm^3
23. 18.4 mL
24. 170 cm^3
25. 412 mL
26. 14.4 cm^3
27. 30.2 mL
28. 164 cm^3
29. 51.7 mL
30. 0.079 L
31. 20 cm^3
32. 130 mL
33. 81 cm^3
34. 4.7 dm^3
35. 44 mL
36. 94.6 cm^3
37. 6.1 L
38. 82 dm^3
39. 707 cm^3
40. 30 L

Group II.

1. 641 kPa
2. 54 mmHg
3. 58 mmHg
4. 430 kPa
5. 663 kPa
6. 855 mmHg

Group III.

1. 210 K
2. 367 K
3. 250 K
4. 314 K
5. 130 K
6. 30.8 K

Group IV.

1. 100 mL
2. 457 mL
3. 226 mL
4. 16 mL
5. 139 mL
6. 102 kPa
7. 15 mL
8. 89 kPa

Group V.

1. 1.73 g/dm^3
2. 0.16 g/dm^3
3. 1.3 g/dm^3
4. 1.7 g/dm^3
5. 1.1 g/dm^3
6. 1.3 g/dm^3
7. 2.6 g/dm^3
8. 1.1 g/dm^3
9. 3.3 g/dm^3
10. 0.72 g/dm^3
11. 0.090 g/dm^3
12. 2.2 g/dm^3
13. 6.4 g/dm^3
14. 8×10^{-5} g/L
15. 890 g/L
16. 1.7 g/L
17. 1 g/L
18. 0.90 g/L
19. 0.090 g/L
20. 2.0 g/L
21. 7 °C
22. 1 700 L
23. 2 800 L
24. Yes
25. 0.88 L
26. Increase of 19 °C
27. 650 mmHg
28. 14 L
29. 110 °C
30. 150 °C

Group VI.

1. 2 300 kPa
2. 81 dm^3
3. 71 amu
4. 227 °C
5. 15.2 kPa
6. 52 amu
7. 814 °C
8. 8.3 mL

Group VII.

1. p_A = 353 mmHg; p_B = 432 mmHg
2. p_A = 0.399 kPa; p_B = 0.472 kPa; p_C = 0.231 kPa
3. p_T = 800 kPa; m_A = 53%; m_B = 23%; m_C = 25%
4. p_B = 0.69 kPa; m_A = 38%; m_B = 62%
5. p_B = 296 kPa; p_C = 392 kPa; p_T = 871 kPa
6. m_{CO2} = 0.227 mol (29.8%); m_{N2} = 0.536 mol (70.2%); p_{CO2} = 0.340 atm; p_{N2} = 0.0802 atm
7. m_{Ar} = 0.125 mol (56.1%); m_{Xe} = 0.038 mol (17.1%); m_{Kr} = 0.060 mol (26.9%); p_{Xe} = 7.1 kPa; p_{Kr} = 41.4 kPa
8. m_{SO2} = 5.3 × 10^{-5} mol (3.5%); m_{CO2} = 4.1 × 10^{-5} mol (2.7%); m_{N2} = 1.4 × 10^{-3} mol (93.8%); p_{SO2} = 0.045 atm; p_{CO2} = 0.035 atm; p_{N2} = 1.20 atm

Group VIII.

1. 1 : 3.74	3. 1 : 4.69	5. 1 : 1.39
2. 1 : 1.17	4. 1 : 1.07	6. 1 : 1.25

Group IX.

1. 2.51 g/L	3. 2.28 g/L	5. 3.88 g/L
2. 5.33 g/L	4. 5.54 g/L	6. 3.34 g/L

Molecular Weight and Mole Calculations (pp. 63–66)

Group I.

1. 98.00	11. 233.00	21. 183.48	31. 126.06
2. 133.34	12. 64.06	22. 223.16	32. 100.95
3. 213.53	13. 474.44	23. 44.01	33. 36.46
4. 226.28	14. 213.90	24. 162.12	34. 399.88
5. 98.08	15. 191.94	25. 63.02	35. 367.00
6. 108.02	16. 368.37	26. 213.01	36. 168.36
7. 244.66	17. 336.48	27. 132.16	37. 187.57
8. 154.76	18. 307.78	28. 244.27	38. 184.11
9. 186.73	19. 212.27	29. 357.49	39. 106.88
10. 149.12	20. 427.62	30. 121.64	40. 96.11

Group II.

1. 119.00 g	6. 53.50 g	11. 54.8 g	16. 230 g
2. 110.98 g	7. 29.17 g	12. 72.1 g	17. 58.6 g
3. 167.96 g	8. 64.45 g	13. 2.4 g	18. 0.39 g
4. 31.01 g	9. 959.92 g	14. 5.15 g	19. 0.25 g
5. 6.90 g	10. 62.47 g	15. 7.56×10^{-3} g	20. 7.97×10^{-4} g

Group III.

1. 1.000 mol	6. 0.333 mol	11. 0.038 mol	16. 7.347×10^{-4} mol
2. 1.00 mol	7. 0.176 mol	12. 0.019 mol	17. 8.18×10^{-7} mol
3. 0.500 mol	8. 0.235 mol	13. 0.124 mol	18. 2.22×10^{-4} mol
4. 2.000 mol	9. 0.0970 mol	14. 0.00102 mol	19. 0.03968 mol
5. 0.25 mol	10. 0.064 mol	15. 0.150 mol	20. 0.04236 mol

Percentage Composition (pp. 67–69)

Answers are given in the same order in which the elements occur in the compound.

1. 39.4%, 60.6%
2. 1.2%, 98.8%
3. 23.5%, 76.5%
4. 42.9%, 57.1%
5. 50.0%, 50.0%
6. 1.5%, 98.5%
7. 71.0%, 29.0%
8. 6.6%, 93.4%
9. 82.4%, 17.6%
10. 14.3%, 4.1%, 81.6%
11. 27.0%, 16.5%, 56.5%
12. 2.0%, 32.6%, 65.4%
13. 24.4%, 17.0%, 58.6%
14. 46.8%, 50.0%, 3.2%
15. 55.2%, 14.6%, 30.2%
16. 50.8%, 16.0%, 33.2%
17. 72.4%, 27.6%
18. 21.2%, 6.1%, 24.3%, 48.4%
19. 25.4%, 12.9%, 57.7%, 4.0%
20. 11.1%, 3.2%, 41.3%, 44.4%
21. 32.8%, 67.2%
22. 78.9%, 21.1%
23. 46.7%, 53.3%
24. 0.8%, 99.2%
25. 27.3%, 72.7%
26. 11.1%, 88.9%
27. 74.3%, 25.7%
28. 11.4%, 88.6%
29. 96.0%, 4.0%
30. 45.9%, 16.5%, 37.6%
31. 26.2%, 7.5%, 66.3%
32. 3.1%, 31.6%, 65.3%
33. 80.1%, 18.7%, 1.2%
34. 12.7%, 19.7%, 67.6%
35. 18.0%, 26.8%, 55.2%
36. 38.8%, 20.0%, 41.2%
37. 53.0%, 47.0%
38. 29.1%, 8.3%, 12.5%, 50.1%
39. 28.1%, 8.2%, 20.8%, 42.9%
40. 56.3%, 29.0%, 1.6%, 13.1%
41. 55.3%
42. 12.0%
43. 36.0%
44. 3.8%
45. 40.5%
46. 32.2%
47. 42.1%
48. 72.7%
49. 35.3%
50. 5.9%
51. 72.5 g
52. 100.0 g
53. 405 kg
54. 3.9 g
55. 9.59 g
56. 8.71 g
57. 173.2 kg
58. 19.54 g
59. 56.8 g
60. yes, 19.2 g are produced.

Empirical Formula (pp. 70–71)

1. Cu_2O	13. CH_3O	25. GaC_3O_6
2. CH_2O	14. CH_4	26. K_2SO_4
3. CH	15. $CaSO_4$	27. HgO
4. Fe_2O_3	16. KNO_3	28. $MgCl_2$
5. HO	17. MgO	29. $BaCl_2$
6. CuO	18. Al_2O_3	30. FeS
7. K_2CO_3	19. Fe_3O_4	31. FeO
8. $CHCl_3$	20. $ZnCO_3$	32. CaO
9. $CrCl_2$	21. As_2S_3	33. CO_2
10. CS_2	22. $HgCl_2$	34. $K_2Cr_2O_7$
11. NO_2	23. $TbCl_3$	35. HgO
12. NH_3	24. Sc_2O_3	

Molecular Formula (pp. 72–74)

Group I.

1. CO_2	6. H_2F_2	11. C_2N_2
2. C_6H_6	7. NO	12. CO
3. H_2S	8. C_2H_4	13. $C_2H_4O_2$
4. S_2Cl_2	9. CH_4O	14. C_2H_6
5. N_2O_4	10. $CHCl_3$	15. $C_{10}H_{14}O_2$

Group II.

1. C_7H_9N	6. $C_4H_6O_2$	11. $C_6H_2O_8N_2$
2. $C_4H_6O_2$	7. $C_4H_6O_4$	12. $C_2H_2O_4$
3. $C_4H_{10}O$	8. $C_6H_8N_2$	13. $C_6H_{12}N_2O_4$
4. $C_{12}H_{24}O_2$	9. $C_{10}H_8S_2O_6$	14. $C_8H_{16}O_4$
5. $C_6H_{12}N_4$	10. $C_{10}H_{14}N_2$	15. $C_6H_{12}O_2$

Soichiometry (pp. 75–85)

- MASS-MASS PROBLEMS

    1. 20.5 g of $CaCO_3$; 9.04 g of CO_2; 22.8 g of $CaCl_2$
    2. 400.0 g of SO_2
    3. 11.7 g of N_2
    4. 78.75 kg of CO; 16.88 kg of C; 105.0 kg of Fe with CO; 105.0 kg of Fe with C
    5. 75.4 g of H_2SO_4; 56.2 g of HCl; 1.54 g of H_2 formed in either case
    6. 500.0 kg of $Ca_3(PO_4)_2$; 561.3 kg of $CaSiO_3$; 290.3 kg of SiO_2
    7. 35 g of H_3PO_4; 72 g of $CaSO_4$
    8. 74.72 g of PH_3; 243.9 g of $Ca(OH)_2$

(continued)

9. 45.45 kg of C; 378.8 kg of As
10. 4.94 g of Fe; 7.18 g of Sb
11. 180 kg of H_3BO_3
12. 10.7 g of Ca
13. 26.8 g of $CaCO_3$
14. 19.8 g of $Ca(OH)_2$
15. 19.8 g of Ca $(OH)_2$
16. 149 g of $MgSO_4$; 176 g of Na_2SO_4
17. 3.7 g of $Ca(HCO_3)_2$
18. 1.8 g of NaOH
19. 0.78 g of Na_2CO_3
20. 37.2 g of $BeCl_2$; 40.5 g of Na_2BeO_2; 34.0 g of HCl; 37.2 g of NaOH
21. 1.67 g of $TbCl_3$
22. 16.8 μg of $GdCl_3$; 17.0 μg of $EuCl_3$; 17.1 μg of $SmCl_3$
23. 13.79 kg of C; 11.84 kg of C; 6.475 kg of C; 4.444 kg of C; 7.947 kg of C
24. 1.8 kg of $LaCl_3$; 1.8 kg of $CeCl_3$
25. 63.94 of $CuSO_4$
26. 3.7 g of Hg; 4.8 g of $SnCl_4$
27. 7.7 mg of $AuCl_3$; 1.6 mg of HNO_3
28. 2.69 kg of C
29. 16.6 g of AgBr
30. 2 177 g of H_2SO_4
31. 21.04 g of NaCl
32. 100 g of KCl (99.78 g)
33. 23.06 g of BaO_2 in xs; 285.3 g of $BaSO_4$
34. 84 g of H_2O; 0.62 g of H_2
35. no; C in excess by 6.2 g
36. 2.3 g of H_2
37. 14.3 g of H_2
38. 11 g of NH_3
39. 44.00 g of CO_2
40. yes
41. yes
42. yes

- MASS-VOLUME PROBLEMS

1. 59 g of FeS
2. 140.0 dm^3 of H_2S; 46.67 dm^3 of H_2S; 200.0 g of S; 46.67 dm^3 of SO_2
3. 14.0 g of MgO; 70.0 g of Mg_3N_2

(continued)

4. 17.5 dm^3 of N_2

5. 710 g of Cu

6. 6.0714 kg of NH_3

7. 4218.8 g of HNO_3

8. 137.1 dm^3 of O_2

9. 0.14 dm^3 of O_2

10. 187 dm^3 of water gas

11. 9.4 g of Li

12. 48.9 g of HCl

13. 40 g of As

14. 43.07 dm^3 of H_2F_2

15. 143.0 dm^3 of H_2

16. 22.2 dm^3

17. 0.4 dm^3 of H_2

18. 161 g of Ca

19. 330.4 of Ca_3N_2

20. 6.89 dm^3 of steam; 6.89 dm^3 of H_2

- VOLUME-VOLUME PROBLEMS

1. 50 dm^3 of H_2; 100 dm^3 of HCl

2. 12 dm^3 of O_2; 25 dm^3 of CO_2

3. 300 dm^3 of H_2; 100 dm^3 of N_2

4. 10 dm^3 of NO; all N_2 is used; 10 dm^3 of O_2 remain

5. H_2 in excess by 20.0 dm^3; 30.0 dm^3 of H_2F_2 formed

6. 25.0 dm^3 of O_2

7. 7 dm^3 per minute

8. 30 per second

- COMBINATIONS

1. 0.44 dm^3 of H_2; 2.4 g of VCl_2

2. 40.0 dm^3 per second; 104 g NaCl per second; 94.6 g Na_2CO_3 per second; 20.0 dm^3 per second of CO_2

3. 78.63 of Na; 38.29 dm^3 of Cl_2

4. 18.4 g of Mg; 17.1 dm^3 of H_2

5. 62.2 dm^3 of H_2; 31.1 dm^3 of O_2

6. 26.019 g of $KClO_3$; 7.1367 dm^3 of O_2

7. 8.0 g of C; 15 dm^3 of CO_2

8. 6.3 dm^3 of Cl_2; 33 g of NaCl

9. 77.7 g of MnO_2; 113 g of $MnCl_2$

10. 0.27 g of Mg; 0.81 g of HCl

11. 2.4 dm^3 of O_2; 42 g of Hg

12. 1.194 g NH_4Cl; 500.0 cm^3 of HCl

Ionic Equations (pp. 86–91)

Group I.

Form: $AB \rightarrow A^+ + B$

Group II.

Form: $A_x^{y+}B_y^{x} \rightarrow {}_xA^{y+} + {}_yB^{x-}$

Group III.

1. H_2S
2. $C_3H_5(OH)_3$
3. $Li^+ Cl^-$
4. $K^+ OH^-$
5. HCN
6. $K_3^+ PO_4^{3-}$
7. $Be^{2+} (OH)_2^{-}$
8. $Ca^{2+} I_2^{-}$
9. HCl
10. $K_2^+ SO_3^{2-}$
11. $NH_4^+ Br^-$
12. CCl_4
13. $Ca^{2+} (OH)_2^{-}$
14. $Sr^{2+} Br_2^{-}$
15. $Na^+ F^-$
16. HI
17. $Na^+ H_2^+ PO_4^{3-}$
18. $Na_2^+ SO_4^{2-}$
19. H_3BO_3
20. $C_{12}H_{22}O_{11}$
21. $Sm^{3+} Br_3^{-}$
22. H_2Se^{3+}
23. $K^+ Br^-$
24. CH_4
25. $NH_4^+ Cl^-$

Group IV.

1. $Na^+Cl^- \rightarrow Na^+ + Cl^-$
2. $Ba^{2+}(OH)_2^{-} \rightarrow Ba^{2+} + 2OH^-$
3. $Ca^{2+} Cl_2^{-} \rightarrow Ca^{2+} + 2Cl^-$
4. $HI \rightarrow H^+ + I^-$
5. $NH_4^+ Br^- \rightarrow NH_4^+ + Br^-$
6. $K_2^+ CO_3^{2-} \rightarrow 2K^+ + CO_3^{2-}$
7. $Na^+ OH^- \rightarrow Na^+ + OH^-$
8. $HC_7H_5O_2 \rightarrow H^+ + C_7H_5O_2^-$
9. no ionization
10. $Na_3^+ PO_4^{3-} \rightarrow 3Na^+ + PO_4^{3-}$
11. $H_2Te \rightarrow 2H^+ + Te^{2-}$
12. $K^+ I^- \rightarrow K^+ + I^-$
13. $Sr^{2+} (OH)_2^{-} \rightarrow Sr^{2+} + 2OH^-$
14. no ionization
15. $HBr \rightarrow H^+ + Br^-$
16. no ionization
17. $Fe^{2+} I_2^{-} \rightarrow Fe^{2+} + 2I^-$
18. $H_2SO_4 \rightarrow 2H^+ + SO_4^{2-}$
19. $Mg^{2+} F_2^{-} \rightarrow Mg^{2+} + 2F^-$
20. $Al^{3+} Cl_3 \rightarrow Al^{3+} + 3Cl^-$
21. $Na_2^+ H^+ PO_4^{3-} \rightarrow 2Na^+ + H^+ + PO_4^{3-}$
22. $HC_2H_3O_2 \rightarrow H^+ + C_2H_3O_2^-$
23. $(NH_4)_2^+ SO_4^{2-} \rightarrow 2NH_4^+ + SO_4^{2-}$
24. $Li^+ F^- \rightarrow Li^+ + F^-$
25. $HCHO_2 \rightarrow H^+ + CHO_2^-$

Group V.

In the following equations, first the ionic, then the net ionic equation, is given. Compounds listed as P in Appendix L are regarded as soluble when they appear as reactants. Those listed as I, A, a, or P are regarded as insoluble when they appear as products.

1. $K^+ + Cl^- + Na^+ + NO_3^- \rightleftharpoons K^+ + NO_3^- + Na^+ + Cl^-$; NR
2. $Rb^+ + OH^- + H^+ + Cl^- \rightarrow Rb^+ + Cl^- + H_2O$; $OH^- + H^+ \rightarrow H_2O$
3. $Li^+ + ClO_3^- + NH_4^+ + Cl^- \rightleftharpoons Li^+ + Cl^- + NH_4^+ + ClO_3^-$; NR
4. $Ca^{2+} + 2 Cl^- + 2 Na^+ + CO_3^{2-} \rightarrow 2 Na^+ + 2 Cl^- + Ca^{2+}CO_3^{2-} \downarrow$; $Ca^{2+} + CO_3^{2-} \rightarrow Ca^{2+}CO_3^{2-} \downarrow$

(continued)

5. $Ca^{2+} + CO_3^{2-} + 2\,Cs^+ + 2\,Cl^- \rightleftharpoons Ca^{2+} + 2\,Cl^- + 2\,Cs^+ + CO_3^{2-}$; NR

6. $2\,NH_4^+ + CO_3^{2-} + 2\,H^+ + 2\,I^- \rightarrow 2\,NH_4^+ + 2\,I^- + H_2CO_3\ (\rightarrow H_2O + CO_2\uparrow)$; $CO_3^{2-} + 2\,H^+ \rightarrow H_2O + CO_2\uparrow$

7. $2\,Al^{3+} + 6\,Br^- + 3\,Cd^{2+} + 6\,NO_3^- \rightleftharpoons 2\,Al^{3+} + 6\,NO_3^- + 3\,Cd^{2+} + 6\,Br^-$; NR

8. NR; CuI_2 not stable in water solution

9. $2\,Fe^{3+} + 3\,SO_4^{2-} + 3\,Pb^{2+} + 6\,ClO_3^- \rightarrow 2\,Fe^{3+} + 6\,ClO_3^- + 3\,Pb^{2+}SO_4^{2-}\downarrow$; $3\,SO_4^{2-} + 3\,Pb^{2+} \rightarrow 3\,Pb^{2+}SO_4^{2-}\downarrow$

10. $2\,Na^+ + 2\,OH^- + 2\,H^+ + SO_4^{2-} \rightarrow 2\,Na^+ + SO_4^{2-} + 2\,H_2O$; $2\,OH^- + 2\,H^+ \rightarrow 2\,H_2O$

11. $Sr^{2+} + 2\,Cl^- + 2\,K^+ + SO_4^{2-} \rightarrow 2\,K^+ + 2\,Cl^- + Sr^{2+}SO_4^{2-}\downarrow$; $Sr^{2+} + SO_4^{2-} \rightarrow Sr^{2+}SO_4^{2-}\downarrow$

12. $2\,Hg^+ + 2\,NO_3^- + Ni^{2+} + SO_4^{2-} \rightarrow Ni^{2+} + 2\,NO_3^- + Hg_2^+SO_4^{2-}\downarrow$; $2\,Hg^+ + SO_4^{2-} \rightarrow Hg_2^+SO_4^{2-}\downarrow$

13. $3\,H^+ + PO_4^{3-} + 3\,NH_4^+ + 3\,OH^- \rightarrow 3\,NH_4^+ + PO_4^{2-} + 3\,H_2O$; $3\,H^+ + 3\,OH^- \rightarrow 3\,H_2O$

14. $K^+ + OH^- + NH_4^+ + Br^- \rightarrow K^+ + Br^- + NH_4OH\ (\rightarrow NH_3\uparrow + H_2O)$; $OH^- + NH_4^+ \rightarrow NH_4OH\ (\rightarrow NH_3\uparrow + H_2O)$

15. $Cu^{2+} + SO_4^{2-} + 2\,Na^+ + 2\,Br^- \rightleftharpoons Cu^{2+} + 2\,Br^- + 2\,Na^+ + SO_4^{2-}$; NR

16. $2\,NH_4^+ + 2\,Br^- + Ca^{2+} + 2\,OH^- \rightarrow Ca^{2+} + 2\,Br^- + 2\,NH_4OH\ (\rightarrow 2\,NH_3\uparrow + 2\,HO_2)$; $2\,NH_4^+ + 2\,OH^- \rightarrow 2\,NH_4OH\ (\rightarrow 2\,NH_3\uparrow + 2\,H_2O)$

17. $Ca^{2+} + 2\,NO_3^- + 2\,K^+ + 2\,Cl^- \rightleftharpoons Ca^{2+} + 2\,Cl^- + 2\,K^+ + 2\,NO_3^-$; NR

18. $2\,Li^+ + 2\,OH^- + 2\,H^+ + S^{2-} \rightarrow 2\,Li^+ + S^{2-} + 2\,H_2O$; $2\,OH^- + 2\,H^+ \rightarrow 2\,H_2O$

19. $2\,Fe^{3+} + 6\,Cl^- + 6\,Na^+ + 3\,S^{2-} \rightarrow 6\,Na^+ + 6\,Cl^- + Fe_2^{3+}S_3^{2-}\downarrow$; $2\,Fe^{3+} + 3\,S^{2-} \rightarrow Fe_2^{3+}S_3^{2-}\downarrow$

20. $3\,Fe^{2+} + 6\,C_2H_3O_2^- + 6\,K^+ + 2\,PO_4^{3-} \rightarrow 6\,K^+ + 6\,C_2H_3O_2^- + Fe_3^{2+}(PO_4)_2^{3-}\downarrow$; $3\,Fe^{2+} + 2\,PO_4^{3-} \rightarrow Fe_3^{2+}(PO_4)_2^{3-}\downarrow$

21. $2\,Na^+ + S^{2-} + 2\,H^+ + 2\,Br^- \rightarrow 2\,Na^+ + 2\,Br^- + H_2S\uparrow$; $S^{2-} + 2\,H^+ \rightarrow H_2S\uparrow$

22. $6\,NH_4^+ + 2\,PO_4^{3-} + 3\,Mg^{2+} + 6\,NO_3^- \rightarrow 6\,NH_4^+ + 6\,NO_3^- + Mg_3^{2+}(PO_4)_2^{3-}\downarrow$; $3\,Mg^{2+} + 2\,PO_4^{3-} \rightarrow Mg_3^{2+}(PO_4)_2^{3-}\downarrow$

23. $Ni^{2+} + 2\,Cl^- + 2\,Na^+ + CO_3^{2-} \rightarrow 2\,Na^+ + 2\,Cl^- + Ni^{2+}CO_3^{2-}\downarrow$; $Ni^{2+} + CO_3^{2-} \rightarrow Ni^{2+}CO_3^{2-}$

24. $Ca^{2+} + 2\,OH^- + 2\,H^+ + 2\,C_2H_3O_2^- \rightarrow Ca^{2+} + 2\,C_2H_3O_2^- + 2\,H_2O$; $2\,OH^- + 2\,H^+ \rightarrow 2\,H_2O$

25. $3\,Zn^{2+} + 6\,Cl^- + 6\,Na^+ + 2\,PO_4^{3-} \rightarrow 6\,Na^+ + 6\,Cl^- + Zn_3^{2+}(PO_4)_2^{3-}\downarrow$; $3\,Zn^{2+} + 2\,PO_4^{3-} \rightarrow Zn_3^{2+}(PO_4)_2^{3-}\downarrow$

26. $3\,Ba^{2+} + 3\,CO_3^{2-} + 6\,H^+ + 2\,PO_4^{3-} \rightarrow Ba_3^{2+}(PO_4)_2^{3-}\downarrow + 3\,H_2CO_3\ (\rightarrow 3\,H_2O + 3\,CO_2\uparrow)$; same

27. $2\,H^+ + 2\,NO_3^- + Sr^{2+} + 2\,OH^- \rightarrow Sr^{2+} + 2\,NO_3^- + 2\,H_2O$; $2\,H^+ + 2\,OH^- \rightarrow 2\,H_2O$

28. $2\,H^+ + SO_4^{2-} + 2\,Na^+ + 2\,Cl^- \rightleftharpoons 2\,H^+ + 2\,Cl^- + 2\,Na^+ + SO_4^{2-}$; NR

29. $2\,K^+ + SO_4^{2-} + 2\,NH_4^+ + 2\,NO_3^- \rightleftharpoons 2\,K^+ + 2\,NO_3^- + 2\,NH_4^+ + SO_4^{2-}$; NR

30. $Pb^{2+} + SO_4^{2-} + 2\,Na + 2\,Br^- \rightleftharpoons Pb^{2+} + 2\,Br^- + 2\,Na^+ + SO_4^{2-}$; NR

31. $Zn^{2+} + 2\,ClO_3^- + 2\,NH_4^+ + S^{2-} \rightarrow 2\,NH_4^+ + 2\,ClO_3^- + Zn^{2+}S^{2-}\downarrow$; $Zn^{2+} + S^{2-} \rightarrow Zn^{2+}S^{2-}\downarrow$

32. $Ca^{2+} + 2\,Cl^- + Mn^{2+} + 2\,I^- \rightleftharpoons Ca^{2+} + 2\,I^- + Mn^{2+} + 2\,Cl^-$; NR

33. $2\,NH_4^+ + SO_4^{2-} + Ba^{2+} + 2\,OH^- \rightarrow Ba^{2+}SO_4^{2-}\downarrow + 2\,NH_4OH\ (\rightarrow 2\,NH_3\uparrow + 2\,H_2O)$; same

34. $K^+ + I^- + NH_4^+ + NO_3^- \rightleftharpoons NH_4^+ + I^- + K^+ + NO_3^-$; NR

35. $3\,Hg^{2+} + 6\,Cl^- + 2\,Al^{3+} + 6\,Br^- \rightleftharpoons 3\,Hg^{2+} + 6\,Br^- + 2\,Al^{3+} + 6\,Cl^-$; NR

(continued)

Group VI.

1. $NH_4^+ + Cl^- + H_2O \rightleftharpoons NH_4OH \ (\rightarrow NH_3 \uparrow + H_2O) + H^+ + Cl^-$
2. $Na^+ + C_2H_3O_2^- + H_2O \rightleftharpoons HC_2H_3O_2 + Na^+ + OH^-$
3. $2 Li^+ + S^{2-} + 2 H_2O \rightleftharpoons Li^+ + 2 OH^- + H_2S\uparrow$
4. $2 NH_4^+ + SO_4^{2-} + 2 H_2O \rightleftharpoons 2 NH_4OH \ (\rightarrow 2 NH_3\uparrow + 2 H_2O) + 2 H^+ + SO_4^{2-}$
5. $K^+ + CN^- + H_2O \rightleftharpoons HCN + K^+ + OH^-$
6. $Rb^+ + C_2H_3O_2^- + H_2O \rightleftharpoons HC_2H_3O_2 + Rb^+ + OH^-$
7. $2 Na^+ + B_4O_7^{2-} + 2 H_2O \rightleftharpoons H_2B_4O_7 + 2 Na^+ + 2 OH^-$
8. $2 NH_4^+ + S^{2-} + 2 H_2O \rightleftharpoons H_2S\uparrow + 2 NH_4OH \ (\rightarrow 2 NH_3\uparrow + 2 H_2O)$
9. $2 Li^+ + CO_3^{2-} + 2 H_2O \rightleftharpoons H_2CO_3 \ (\rightarrow H_2O + CO_2\uparrow) + 2 Li^+ + 2 OH^-$
10. $NH_4^+ + NO_3^- + H_2O \rightleftharpoons NH_4OH \ (\rightarrow NH_3\uparrow + H_2O) + H^+ + NO_3^-$
11. $Ca^{2+} + 2 C_2H_3O_2^- + H_2O \rightleftharpoons 2 HC_2H_3O_2 + Ca^{2+} + 2 OH^-$
12. $2 Na^+ + CO_3^{2-} + 2 H_2O \rightleftharpoons H_2CO_3 \ (\rightarrow H_2O + CO_2\uparrow) + 2 Na^+ + 2 OH^-$
13. $NH_4^+ + C_2H_3O_2^- + H_2O \rightleftharpoons HC_2H_3O_2 + NH_4OH \ (\rightarrow NH_3\uparrow + H_2O)$
14. $2 K^+ + S^{2-} + 2 H_2O \rightleftharpoons H_2S\uparrow + 2 K^+ + 2 OH^-$
15. $NH_4^+ + HCO_3^- + H_2O \rightleftharpoons H_2CO_3 \ (\rightarrow H_2O + CO_2\uparrow) + NH_4OH \ (\rightarrow NH_3\uparrow + H_2O)$

Equilibria (pp. 92–96)

Group I.

1. $K_{eq} = \dfrac{[SO_3]^2}{[SO_2]^2 \times [O_2]}$

2. $K_{eq} = \dfrac{[CO_2]^2}{[CO]^2 \times [O_2]}$

3. $K_{eq} = \dfrac{[NH_3]^2}{[N_2] \times [H_2]^3}$

4. $K_{eq} = \dfrac{[HCl]^2}{[H_2] \times [Cl_2]}$

5. $K_{eq} = \dfrac{[N_2O]^2}{[N_2]^2 \times [O_2]}$

6. $K_{eq} = \dfrac{[NO_2]^2}{[NO]^2 \times [O_2]}$

7. $K_i = \dfrac{[H^+] \times [C_2H_3O_2^-]}{[HC_2H_3O_2]}$

8. $K_i = \dfrac{[H^+] \times [CN^-]}{[HCN]}$

9. $K_{sp} = [Ag^+] \times [Cl^-]$ ([AgCl] = constant)
10. $K_{sp} = [Pb^{2+}] \times [I^-]^2$ ([PbI_2] = constant)
11. $K_{sp} = [Bi^{3+}]^2 \times [S^{2-}]^3$ ([Bi_2S_3] = constant)
12. $K_{sp} = [Ca^{2+}]^3 \times [PO_4^{3-}]^2$ ([Ca_3(PO_4)_2] = constant)

(continued)

Group II.

1. 4.0
2. 50
3. 0.4
4. 0.045
5. 0.022
6. 0.2
7. 2.67×10^4
8. 0.116

9. 1.0×10^{-5}
10. 2×10^{-7}
11. 1.35×10^{-4}
12. 1.44×10^{-5}
13. 9.12×10^{-14}
14. 6.16×10^{-30}
15. 1.23×10^{-3} mol/L
16. 1.57×10^{-5} g/mL

Group III.

1. $K = \dfrac{[C]^3 [D]}{[A]^2 [B]} = 1.78$
2. $[A] = [B] = 0.071$ M
3. $K = 0.058$ mol/dm^3
4. $K = 1.59 \times 10^{-4}$
5. $[H_2] = 0.10$
6. $[O_2] = 0.5$

7. $K = 85.2$
8. $K_D = 1$
9. $K_{sp} = 20 \times 10^{-12}$
10. 5×10^{-8} mol/dm^3
11. $K_{sp} = 1.58 \times 10^{-12}$
12. 3×10^{-39} mol/dm^3
13. $K_{sp} = 9.42 \times 10^{-10}$

14. Yes, AgBr will precipitate out ($8.28 \times 10^{-9} > 4.9 \times 10^{-13}$)
15. $[Ca^{2+}] = 7.43 \times 10^{-6}$ M; $[PO_4^{-3}] = 4.95 \times 10^{-6}$ M
16. $[Pb^{2+}] = 1.78 \times 10^{-2}$ mol/dm^3

17. **Le Chatelier's Principle**

    The effect of (a) increased pressure and (b) increased temperature on each reaction is indicated below by an arrow showing the reaction favored in each case.

Reaction	Increased Pressure	Increased Temperature
1	←	→
2	←	←
3	←	→
4	→	neither
5	→	←
6	neither	→

Acids and Bases (pp. 97–100)

Group I.

1. 10^{-1}, 10^{-13}, 1, 13
2. 1, 10^{-14}, 0, 14
3. 10^{-13}, 10^{-1}, 13, 1
4. 10^{-14}, 1, 14, 0
5. 10^{-3}, 10^{-11}, 3, 11

(continued)

6. 10^{-3}, 10^{-11}, 3, 11

7. 10^{-3}, 10^{-11}, 3, 11

8. 10^{-11}, 10^{-3}, 11, 3

9. 10^{-10}, 10^{-4}, 10, 4

10. 10^{-10}, 10^{-4}, 10, 4

11. 5×10^{-4}, 2×10^{-11}, 3.3, 10.7

12. 1.5×10^{-4}, 6.7×10^{-11}, 3.8, 10.2

13. 3.5×10^{-6}, 2.9×10^{-9}, 5.5, 8.5

14. 1.8×10^{-4}, 5.5×10^{-11}, 3.7, 10.3

15. 8.7×10^{-5}, 1.1×10^{-10}, 4.1, 9.9

16. 6.7×10^{-7}, 1.5×10^{-8}, 6.2, 7.8

17. 3.2×10^{-8}, 3.1×10^{-7}, 7.5, 6.5

18. 2.7×10^{-6}, 3.8×10^{-9}, 5.6, 8.4

19. 5.0×10^{-5}, 2.0×10^{-10}, 4.3, 9.7

20. 3.5×10^{-6}, 2.8×10^{-9}, 5.5, 8.5

21. 1.3×10^{-12}, 7.7×10^{-3}, 11.9, 2.1

22. 6.4×10^{-12}, 1.6×10^{-3}, 11.2, 2.8

23. 7.9×10^{-11}, 1.3×10^{-4}, 10.1, 3.9

24. 1.2×10^{-12}, 8.4×10^{-3}, 11.9, 2.1

25. $[H^+] = 13.2 \times 10^{-5}$

26. pH = 2.44

27. $K_i = 1.9 \times 10^{-6}$

28. (a) 0.50 M CH_3COOH (b) $[H^+] = 3.0 \times 10^{-3}$

29. $[H^+] = [ClO^-] = 4.10 \times 10^{-5}$ M

30. pH = 3.0

31. pH = 9.05

32. (a) $HF(aq) + CH_3COO^-(aq) \rightarrow F^-(aq) + CH_3COOH(aq)$

  (b) $H_2S(aq) + CO_3^{-2}(aq) \rightarrow HS^- + HCO_3^-(aq)$

  (c) $NH_4^+(aq) + F^-(aq) \rightarrow HF(aq) + NH_3(aq)$

33. (a) $K_a = \dfrac{[NO_2^-][NH_4^+]}{[HNO_2][NH_3]}$ (b) $K_a = \dfrac{[C_6H_5COO^-][CH_3COOH]}{[C_6H_5COOH][CH_3COO^-]}$

34. (a) HSO_4^- and SO_4^{2-}; $C_2O_4^{2-}$ and $HC_2O_2^-$;   (b) $H_2PO_4^-$ and HPO_4^{2-}; HCO_3^- and H_2CO_3;   (c) H_2O and OH^-, S^{2-} and HS^-;   (d) HCN and CN^-; $HC_2H_3O_2$ and $C_2H_3O_2^-$;   (e) HNO_2 and NO_2^-; H_3O^+ H_2O.

35. (a) base; (b) base; (c) base; (d) acid; (e) acid; (f) acid; (g) base; (h) acid

Group II.

1. [HClO] = 0.25 M
2. [KOH] = 0.15 M
3. [HClO] = 0.18 M
4. [Ca(OH)$_2$] = 0.22 M

5. [Ca(OH)$_2$] = 0.30 M
6. [NaOH] = 0.3 M
7. [HCN] = 0.23 M
8. [H$_2$SO$_4$] = 0.10 M

9. [Ca(OH)$_2$] = 0.12 M
10. [HClO$_4$] = 0.18 M

Standard Solutions (pp. 101–111)

• STANDARD SOLUTIONS

Group IA.

1. 74.6 g in 1 dm^3
2. 211.6 g in 1 L
3. 142 g in 1 dm^3
4. 173.7 g in 2 L
5. 1065 g in 5 dm^3
6. 11.1 g in 1 L
7. 13.3 g in 1 dm^3
8. 17 g in 1 L
9. 0.58 g in 1 dm^3
10. 4.8 g in 500 mL
11. 8.36 g in 40 cm^3
12. 4.68 g in 600 mL
13. 6.84 g in 100 cm^3
14. 0.54 g in 250 mL
15. 53 g in 50 cm^3
16. 10.45 g in 400 mL
17. 492 g in 500 cm^3
18. 1.95 g in 25 mL
19. 18.37 g in 100 cm^3
20. 738 g in 750 mL
21. 36.5 g in 1 L
22. 147 g in 1 dm^3
23. 44.4 g in 1 L
24. 32.7 g
25. 255 g
26. 80. g in 1 L
27. 129.6 g in 1 L
28. 284 g in 1 L
29. 378 g in 1 L
30. 6.8 g in 1 dm^3

31. 8.6 g in 1 dm^3
32. 10.42 g in 1 dm^3
33. 8.1 g in 1 dm^3
34. 22.3 g in 1 dm^3
35. 14.8 g in 1 L
36. 1.025 g in 1 L
37. 0.91 g in 1 L
38. 0.608 g in 1 dm^3
39. 2.99 g in 1 dm^3
40. 6.8 g in 1 dm^3
41. 114 g in 2 L
42. 375.1 g in 4 L
43. 64 g in 500 cm^3
44. 12.7 g in 200 cm^3
45. 42 g in 750 cm^3
46. 13.57 g in 100 cm^3
47. 69.8 g in 250 cm^3
48. 1.47 g in 300 mL
49. 2.77 g in 500 mL
50. 0.774 g in 100 mL
51. 1.75 g in 250 mL
52. 16.1 g in 50 mL
53. 2.72 g in 25 cm^3
54. 0.049 g in 30 cm^3
55. 2.40 g in 200 cm^3
56. 0.50 g in 350 cm^3
57. 0.239 g in 150 mL
58. 13.1 g in 50 mL
59. 3.8 g in 250 mL
60. 94.57 g in 8 L

Group IB.

1. 1.0; 1.0
2. 1.000; 1.000
3. 1.000; 3.000
4. 1.000; 2.000
5. 1.00; 6.00
6. 1.00; 1.00
7. 2.00; 4.00
8. 0.125; 0.125
9. 0.28; 0.28
10. 1.33; 4.00
11. 025; 0.25
12. 0.50; 0.50
13. 0.324; 0.648
14. 0.635; 0.635
15. 0.900; 5.40
16. 0.044; 0.044
17. 0.11; 0.22
18. 0.324; 0.972
19. 0.77; 1.5
20. 0.557; 1.11
21. 0.0085; 0.017
22. 0.77; 0.77
23. 0.62; 0.62
24. 0.026; 0.17
25. 0.087; 0.17

• PERCENT SOLUTIONS

Group II.

1. 5.0% (m/v)
2. 2.5% (m/m)
3. 1.0% (v/v)
4. 6.0% (m/v)
5. 1.0% (m/v)
6. 3.0% (m/m)
7. 7.5% (v/v)
8. 6.3% (m/v)
9. 1.38% (m/m)
10. 0.577% (v/v)
11. 3.2% (m/v)
12. 0.217% (m/m)

Group III.

1. 1.5 g
2. 2.3 g
3. 1.3 g
4. 0.063 g
5. 0.90 g
6. 1.4 g
7. 0.59 g
8. 2.0 g
9. 0.79 g
10. 0.050 g
11. 0.0048 g
12. 1.02 g

Group IV.

1. 100 mL
2. 200 mL
3. 50 mL
4. 300 mL
5. 12.5 mL
6. 111.1 mL
7. 560 mL
8. 130 cm^3
9. 400 mL
10. 230 mL
11. 560 mL
12. 320 cm^3

Group V.

1. 12
2. 15.6
3. 36.1
4. 15
5. 17.6
6. 2.7
7. 8
8. 5.3
9. 5.30
10. 20

Group VI.

Dilute each of the given solutions with distilled water as follows.

1. to 1.0 L
2. to 6.0 L
3. to 5.0 L
4. to 2.0 L
5. to 2.4 L
6. to 3.0 L
7. to 10 L
8. to $5\overline{0}0$ mL
9. to $4\overline{0}0$ mL
10. to 1.0 L

Group VII.

Dilute each of the given solutions with distilled water as follows.

1. to 5.0 L
2. to $2\overline{0}0$ L
3. to 1800 L
4. to $3\overline{0}0$ L

5. to $5\overline{0}0$ mL
6. to 25 L
7. to 1.0 L

8. to $2\overline{0}0$ L
9. to 10. L
10. to 40. L

- TITRATIONS

Group VIII.

1. (a) 20. mL
   (b) $4\overline{0}0$ mL
   (c) 1.0 mL
   (d) 0.80 mL
   (e) 2.0 L
   (f) 4.0 mL
   (g) 10. mL
   (h) 2.0 mL
   (i) 40. mL
   (j) 8.0 mL
2. (a) 1.0 N
   (b) 0.040 N
   (c) 2.0 N
   (d) 0.10 N

   (e) 0.25 N
   (f) 0.14 N
   (g) 0.50 N
   (h) 0.16 N
   (i) 0.80 N
   (j) 0.33 N
3. 40. cm^3
4. 0.48 N
5. 0.29 N
6. 4.5 mL
7. 6.0 cm^3
8. 1.3 N
9. 0.0077 N
10. 79 cm^3

11. (a) 1250 mL
    (b) 27 mL
    (c) 2200 mL
    (d) 830 mL
    (e) 1.2 mL
    (f) 7500 mL
    (g) 1100 mL
    (h) 1800 mL
    (i) 270 mL
    (j) 61 mL
12. 0.482 N
13. 0.662 N
14. 270 mL

15. 0.49 N
16. (a) 0.946 N
    (b) 0.289 N
    (c) 0.342 N
    (d) 0.122 N
    (e) 0.899 N
    (f) 0.218 N
    (g) 0.520 N
    (h) 0.351 N
    (i) 0.462 N
    (j) 0.458 N
    (k) 0.295 N
    (l) 0.0299 N

- MOLAL SOLUTIONS

Group IX.

1. (a) Dissolve 58.5 g of NaCl in 1.0 kg of water.

   (b) Dissolve 200 g (196 g) of H_2SO_4 in 1 kg of water.

   (c) Dissolve 86 g of sucrose in 1 kg of water.

   (d) Dissolve 23 g of ethyl alcohol in 500 g of water.

   (e) Dissolve 24.5 g of H_3PO_4 in 500 g of water.

   (f) Dissolve 24 g of HNO_3 in 1.5 kg of water.

   (g) Dissolve 0.90 g of NaOH in 75 g of water.

   (h) Dissolve 2.62 g of NH_4OH in 150 g of water.

2. (a) 1.00 m;
   (b) 0.250 m;

   (c) 0.342 m;
   (d) 0.702 m;

   (e) 0.17 m;
   (f) 0.12 m;

   (g) 1.22 m;
   (h) 1.2 m

Colligative Properties (pp. 112)

1. 100.52 °C
2. –0.93 °C
3. 100.15 °C
4. –0.640 °C
5. 2.43 °C
6. 35
7. 62.6
8. 140
9. 58.5
10. 37.5 g

Redox Reactions (pp. 113–117)

Given below are the coefficients that the equations will have when properly balanced. They are given in the same order as that found in the original problem.

1. 1-1-1-1-2
2. 2-1-2-1
3. 1-2-1-2-1
4. 1-1-2-1-2-1
5. 1-6-3-1-3-3
6. 2-4-2-2-1
7. 1-2-1-2-1
8. 1-2-1-2
9. 1-1-2-2
10. 4-1-4-1
11. 10-2-48-15-1-2-48
12. 2-2-3-2-4
13. 3-4-3-1-2
14. 3-8-6-2-3-4
15. 1-4-1-1-1-2
16. 3-3-2-1-3
17. 6-6-5-1-6
18. 1-2-2-1-1-2
19. 1-2-1-2
20. 1-9-9-1-3-9
21. 1-10-10-2-2-10-7
22. 3-3-1-3-1-2
23. 2-16-2-2-3-8
24. 1-8-3-2-2-3-7
25. 4-4-1-1-4-2-2
26. 3-6-2-3-3-2-4
27. 1-2-7-2-2-5
28. 3-4-2-1-4-2
29. 2-5-6-2-2-5-3
30. 1-4-1-1-2-2-2
31. 1-2-1-1-1-2
32. 2-5-3-2-5-4
33. 2-5-26-2-15-12
34. 4-8-4-6-5
35. 3-6-4-4-3-6-4
36. 1-2-2-1-1
37. 3-5-3-5-1-5
38. 3-5-1-3-3-3-3
39. 5-2-2-2-2-5-2
40. 3-12-14-6-6-14-10
41. 1-1-1-1-1
42. 2-16-5-2-5-5-7
43. 2-15-1-4-11
44. 1-2-1-1-1
45. 6-5-2-10-6-5-2-2-1
46. 3-5-1-3-3-6-3
47. 1-2-2-1-2-1-1
48. 1-10-12-1-10-6-6
49. 1-1-2-1-2-1
50. 1-2-2-1-2-1-1
51. 3-4-1-3-4
52. 4-5-12-4-8-5-6
53. 2-1-1-1-2
54. 3-30-10-6-15-10-20
55. 2-16-2-2-5-8
56. 3-5-2-2-5
57. 3-10-4-6-15-10
58. 1-6-2-2-3
59. 1-1-2-1-1-1
60. 2-3-3-1-3-3
61. 1-3-1-3
62. 1-1-1-1
63. 15-20-4-12-3-12-12
64. 2-4-1-2-2
65. 2-4-4-2-4-1-2
66. 1-4-6-6-1-2-3
67. 1-1-2-1-2
68. 1-5-6-6-3-3
69. 6-1-1-3-3
70. 1-1-14-6-2-2-7
71. 5-4-24-10-2-4-9
72. 1-6-1-6
73. 5-6-16-10-6-3-6
74. 1-2-2-2-4
75. 6-1-5-6-3-3
76. 1-5-1-3-2-3-3
77. 1-2-2-1-1-1
78. 2-3-3-2-4

Nuclear Chemistry (pp. 118–122)

Group I.

1. $^{63}_{30}Zn$

2. $^{1}_{0}n$

3. $^{1}_{0}n$

4. $^{1}_{1}H$

5. $^{2}_{1}H$

6. $^{3}_{1}H$

7. $^{1}_{0}n$

8. $^{1}_{0}n$

9. γ

10. α

11. $^{242}_{96}Cm$

12. $^{25}_{13}Al$

13. $^{1}_{1}H$

14. $^{139}_{57}La$

15. $^{1}_{0}n$

16. $^{3}_{1}H$

17. γ

18. $^{214}_{83}Bi$

19. $^{62}_{29}Cu$

20. $^{1}_{0}n$

21. $^{241}_{95}Am$

22. $^{1}_{0}n$

23. $^{1}_{1}H$

24. $^{1}_{1}H$

25. γ

26. $^{1}_{1}H$

27. α

28. $^{30}_{15}P$

29. $X = \alpha$

Group II.

Missing member only is shown.

1. $^{7}_{4}Be$

2. $^{1}_{1}H$

3. $^{1}_{1}H$

4. $^{13}_{7}N$

5. $^{125}_{53}I$

6. α

7. $^{27}_{12}Mg$

8. $^{7}_{4}Be$

Group III.

Various methods are possible in each case. Only one possible method is shown here.

1. $^{152}Sm(d, \gamma)^{154}Eu$

2. $^{19}F(p,d)^{18}F$

3. $^{45}Sc(n, \gamma)^{46}Sc$

4. $^{94}Zr(p, \gamma)^{95}Nb$

5. $^{84}Kr(d,p)^{85}Kr$

6. $^{142}Ce(p,n)^{142}Pr$

7. $^{88}Sr(d, \gamma)^{90}Y$

8. $^{100}Ru(\alpha, n)^{103}Pd$

Group IV.

1. 4.0 g
2. 2.0 g
3. 4.0 g
4. 1.56 g
5. 1.88 g
6. 40.8 d
7. 13,950 y
8. 16.0 g
9. 80. g
10. 3.0 m

Organic Chemistry (pp. 123–140)

Group I.

1. alkane (none)
2. alcohol (–OH)
3. aldehyde (–CHO)
4. alkane (none)
5. aldehyde (–CHO)

6. ether (–O–)
7. ketone (–C=O)
    |
8. alkene (–C=C–)
    | |

9. alkyl halide (halogenated hydrocarbon) (–X)
10. carboxylic acid (–COOH)
11. alkyne (–C≡C–)

Group II.

1. n-hexane
2. 3-methylheptane
3. 3-ethyl-2,3-dimethylhexane
4. n-butane
5. 3-ethylpentane
6. 5-ethyl-2-methylheptane
7. 2,3,4-trimethylheptane
8. 2-bromo-4-methylhexane
9. 5-ethyl-3,4-dimethylheptane
10. 5,6-dichloro-2-methylnonane
11. 2,7-dibromo-1-chloro-4-methyloctane
12. 7-chloro-5-ethyl-2-methyl-4-propyloctane
13. 3-bromo-5-ethylnonane
14. 1,4,6-triiodooctane
15. 6-fluoro-3-iodo-4-methyloctane
16. 4-ethyl-6-iodo-2-methylnonane
17. ethene
18. 3-heptene
19. 2,3-dimethyl-3-hexene
20. 5-chloro-4-methyl-1-hexene
21. 6-bromo-4-methyl-2-octene
22. 5,6-dimethyl-2-heptene
23. 7-chloro-4-ethyl-2-methyl-3-heptene
24. 6-chloro-4,4,7,7-tetramethyl-2-nonene
25. 7-bromo-5,5-dichloro-2-heptene
26. 4-ethyl-3-heptene
27. 2,4-dibromo-6-methyl-3-heptene
28. 5,7-dibromo-4-methyl-3-heptene
29. 5-bromo-7,8-dichloro-3-octene
30. propyne
31. 2-butyne
32. 1-bromo-2-pentyne
33. 5-methyl-2-hexyne
34. 2-bromo-3-methyl-4-octyne
35. 1-propanol
36. 2-pentanol
37. 2-butanol
38. 4-heptanol
39. 2-heptanol
40. 2-chloro-3-pentanol
41. 2-methyl-1-butanol
42. 3-methyl-2-pentanol
43. 3,5-dimethyl-2-hexanol
44. 2,2,6-trimethyl-4-heptanol
45. 4-chloro-5-methyl-1-hexanol
46. 1,2-dichloro-4-methyl-3-pentanol
47. 2,3-dimethyl-4-heptanol
48. 1,3-dichloro-2-butanol
49. 5,5-diethyl-2-heptanol
50. 6-bromo-4,7-dichloro-2-heptanol
51. 3-ethyl-2-methyl-1-hexanol
52. 1,4-butanediol
53. 1,4-hexanediol
54. ethanal
55. 3-hexanone
56. butanal
57. 3-methylbutanal
58. 3,3-dimethylpentanal
59. 3-chloropentanal
60. 2,4-4-trimethylpentanal
61. 4,4-dimethylpentanal
62. 3-octanone
63. 2-methyl-4-heptanone
64. 5,5-dimethyl-3-hexanone
65. 3,6-dichloroheptanal
66. 5-iodo-7-methyl-2-octanone
67. 2,5-diiodohexanal
68. 4-iodo-6,6-dimethyl-3-heptanone
69. 1-bromo-7-methyl-4-octanone
70. 5,5,7-trichloro-3-heptanone
71. 3-methylhexanal
72. 1-chloro-7,7-dimethyl-4-octanone
73. 3,3-dimethyl-5-chloroheptanal
74. dimethyl ether
75. methylethyl ether
76. methyl-n-propyl ether
77. methyl-n-pentyl ether
78. ethyl-t-butyl ether

(continued)

79. butanoic acid
80. pentanoic acid
81. 2,2-dimethylbutanoic acid
82. 3-methylbutanoic acid
83. butanedioic acid
84. 4-methylhexanoic acid
85. 2,4-dimethylpentanoic acid
86. 3-chlorohexanoic acid

87. 4-fluorohexanoic acid
88. 2-aminobutanoic acid
89. 2-bromobutanoic acid
90. 2,4-dibromopentanoic acid
91. 2-aminopentanoic acid
92. 3-hydroxybutanoic acid
93. 2-hydroxyhexanoic acid

Group III.

1. $CH_3CH_2CH_2CH_3$
2. $CH_3CHClCH_2CH_2CH_3$
3. $CH_3CH_2CH(CH_3)CH_2CH_2CH_3$
4. $CH_3(CH_2)_5CH_3$
5. $CH_2ClCHClCH_2CHClCH_3$
6. $CH_3CH(CH_3)CH(CH_3)CH(CH_3)CH_2CH_3$
7. $CH_3CH(CH_3)C(CH_3)_2CH_2CH_2CH_2CH_3$
8. $CH_3CH(CH_3)CH_2CClBr(CH_2)_5CH_3$
9. $CH_3CH(CH_3)C(CH_3) = CHCH_2CH_3$
10. $CH_2 = CClCH_2CH_2CH_3$
11. $CH_3C(CH_3)_2C(CH_3) = CHCH(CH_3)CH_2CH_2CH_3$
12. $CH_2BrCCl_2CH = CBrCHBrCH_2CH_3$
13. $CH_2 = CFCClFCH_3$
14. $CH_2 = CHCH = CH_2$
15. CH_3CH_2OH
16. $CH_3CHOHCH_3$
17. $CH_3CHOHCH_2CH_2CH_3$
18. $CH_3CHOHC(CH_3)_2CH_2CH_2CH_3$
19. $CHClOHCHClCHClCH_3$
20. $CH_2OHC(CH_3)_2CH(C_2H_5)CH_2CH_2CH_3$
21. $CH_3C(CH_3)(OH)CH_2C(CH_3)_2CH_2CH_2CH_3$
22. $CHClOHCHBrCHClC(CH_3)_2CH_2CHClCHClCH_2CH_2CH_3$
23. $CH_2OHCH_2CHOHCH_3$
24. $CH_3CBr_2C(CH_3)(OH)CH(CH_3)CH_2CH_3$
25. $CHOHCl(CH_3)CH(C_2H_5)Cl(CH_3)$
26. $CH_3(CH_2)_5CHO$
27. $CH_3CH_2CH_2CHO$

28. $CH_3\overset{*}{C}OCH_2CH_2CH_3$
29. $(C_2H_5)_2\overset{*}{C}O$
30. $CH_3CH_2CH_2\overset{*}{C}OCH_2CH_2CH_2CH_2CH_3$
31. $CH_3CH_2CH_2CH(C_2H_5)CH(CH_3)CHO$
32. $CH_3CHBrCCl_2CHO$
33. $CH_3CCl(CH_3)\overset{*}{C}OCH(CH_3)CH_2CH_3$
34. CH_2O
35. $CH_3CH_2C(CH_3)_2CH(CH_3)CHO$
36. $CH_3\overset{*}{C}OC(CH_3)_2CClICH_3$
37. $CH_3CH_2CH_2CHClCHClCH_2CHClCHBrCHBrCHO$
38. $CH_3CH(CH_3)CH(CH_3)\overset{*}{C}OC(CH_3)(C_2H_5)CH_2CH(C_2H_5)CH_2CH_3$
39. $CH_3CH_2CH_2COOH$
40. $HOOCCH_2CH_2CH_2COOH$
41. $CH_3CH(NH_2)COOH$
42. $CH_3CHClCHClCOOH$
43. $CH_3CH_2CH_2C(CH_3)_2CH_2COOH$
44. $CH_3CHOHCH_2CH_2COOH$
45. $CH_3(CH_2)_5CHBrCHBrCHBrCOOH$
46. $CH_3CHCl_2CH_2CH(CH_3)COOH$
47. $CH_3CH_2CHOHCOOH$
48. $CH_3CH_2CHOHCH_2COOH$
49. $CH_3CH_2CHClCH(CH_3)CH(CH_3)COOH$
50. $CH_3CH_2CH_2CHClCH_2CHClCHBrCHBrCHClCOOH$
51. $CH_3CH_2C(NH_2)_2COOH$
52. $CH_3CH_2CHClCHClCH_2CH_2CH_2COOH$
53. $CH_3CHOHCH_2CHOHCOOH$

* indicates a ketone group, not an ether linkage

Group IV.

1. $CH_4 + Cl_2 \rightarrow CH_3Cl(+Cl_2) \rightarrow CH_2Cl_2(+Cl_2) \rightarrow CHCl_3(+Cl_2) \rightarrow CCl_4$

2. $CH_3CH = CH_2 + Cl_2 \rightarrow CH_3CHClCH_2Cl$

3. $C_2H_5Cl + NaOH \rightarrow NaCl + C_2H_5OH$

4. $C_2H_6 + Br_2 \rightarrow C_2H_5Br + HBr$

5. $CH_3CH_2CH = CHCH_2CH_3 + HCl \rightarrow CH_3CH_2CH_2CHClCH_2CH_3$

6. $CH_3CH = CHCH_3 + H_2 \rightarrow CH_3CH_2CH_2CH_3$

7. $C_6H_6 + Cl_2 \rightarrow C_6H_5Cl + HCl$

8. $2\ C_8H_{18} + 25\ O_2 \rightarrow 16\ CO_2 + 18\ H_2O$

9.
$$\overset{\overset{\displaystyle O}{\|}}{CH_3C}\ CH_2CH_2CH_3 + HCN \rightarrow CH_3C(OH)\ (CN)CH_2CH_2CH_3$$

10. $2CH_2OHCH_2CH_2CH_3 + 2Na \rightarrow 2CH_2ONaCH_2CH_2CH_3 + H_2$

11. $CH_3CH_2CH_2CH_3 + Br_2 \rightarrow$ mixture of $CH_2BrCH_2CH_2CH_3 + CH_3CHBrCH_2CH_3$

12. $CH \equiv CH + HCl \rightarrow CH2 = CHCl$

13. $CH_3CH = CH_2$ (polymerized) $\rightarrow$ (first step) $CH_3CH \diagdown \genfrac{}{}{0pt}{}{CH_3}{CH_2CH = CH_2} \rightarrow$

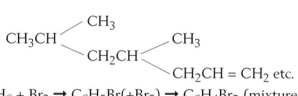

14. $C_6H_6 + Br_2 \rightarrow C_6H_5Br(+Br_2) \rightarrow C_6H_4Br_2$ (mixture of ortho and para forms)

15. $HCOOH + CH_3CH_2CHOHCH_2CH_2CH_3 \rightarrow CH_3CH_2CH(OOCH)CH_2CH_2CH_3$
    $+ H_2O$

16. $CH_2 = CHCH_2CH_2CH_3 + HBr \rightarrow CH_3CHBrCH_2CH_2CH_3$

17. $2\ C_{12}H_{26} + 37\ O_2 \rightarrow 24\ CO_2 + 26\ H_2O$

18. $C_6H_5CH_3 + Br_2 \rightarrow C_6H_5CH_2Br$(or $C_6H_4ClCH_3$) $+ HBr$

19. $CH_3CH = CH(CH_2)_4CH_3 + CH_3CH_2CH_2CH_3 \rightarrow CH_3CH_2CH(CH_2CH_2CH_2CH_3)$
    $(CH_2)_4CH_3$

20. $CH_3CH_2CH_3 + Cl_2 \rightarrow$ mixture of $CH_2ClCH_2CH_3$ and $CH_3CHClCH_3$

21. $CH_2 = CHCH_2CH_3 + H_2 \rightarrow CH_3CH_2CH_2CH_3$

22. $CH_3CH = CHCH_2CH_3 + HCN \rightarrow$ mixture of $CH_3CH_2CH(CN)CH_2CH_3$ and
    $CH_3CHCNCH_2CH_2CH_3$

23.
$$CH_3CHOHCH_2CH_3 + (O) \rightarrow \overset{\overset{\displaystyle O}{\|}}{CH_3CCH_2CH_3} + H_2O$$

24. $CH_3CH_2CH = CH_2$ (polymerized) $\rightarrow$ (step 1) $CH_3CH_2CH(CH_2CH_2CH = CH_2)CH_3$, etc.

25. $CH \equiv CCH_3 + Br_2 \rightarrow CHBr = CBrCH_3$

26. $CH_3CH_2CH_2CH_2CH_3 + Br_2 \rightarrow$ 1,2, or 3 monobromopentane $+ HBr$

27. $2CH_3CHOHCH_2CH_2CH_3 + 2Na \rightarrow 2CH_3CHONaCH_2CH_2CH_3 + H_2$

(continued)

28. $CH_3CH = CHCH_2CH_2CH_3 + CH_3CH_2CH_3 \rightarrow$ mixture of
$CH_3CH_2CH(CH_2CH_2CH_3)CH_2CH_2CH_3$ and $CH_3CH(CH_2CH_2CH_3)CH_2CH_2CH_2CH_3$

29. $CH_3CH = CHCH_2CH_3 + Br_2 \rightarrow CH_3CHBrCHBrCH_2CH_3$

30. $3CH_3CH_2CHOHCH_2CH_2CH_3 + PCl_3 \rightarrow 3CH_3CH_2CHClCH_2CH_2CH_3 + P(OH)_3$

31. $2 C_{18}CH_{38} + 55 O_2 \rightarrow 36 CO_2 + 38 H_2O$

32. $CH_2 = CHCH_2CH_3 + CH_3CH_3 \rightarrow CH_3CH(CH_2CH_3)CH_2CH_3$

33. $CH_3CH_2CH_2CH_2COOH + CH_3CH_2CH_2CH_2OH \rightarrow$
$CH_3CH_2CH_2CH_2COOCH_2CH_2CH_2CH_3 + H_2O$

34.
$$CH_3CH_2\overset{\overset{\displaystyle O}{\|}}{C} CH_2CH_2CH_3 + HCN \rightarrow CH_3CH_2C(OH)(CN)CH_2CH_2CH_3$$

35.
$$CH_3CH_2CHOHCH_2CH_2CH_3 + (O) \rightarrow CH_3CH_2\overset{\overset{\displaystyle O}{\|}}{C} CH_2CH_2CH_3$$

36. $(C_{15}H_{31}COO)(C_{17}H_{35}COO)_2C_3H_5 + 3HOH \rightarrow C_3H_5(OH)_3 + C_{15}H_{31}COOH + 2C_{17}H_{35}COOH$

37. $CH_3CH_2OH + CH_3COOH \rightarrow CH_3COOCH_2CH_3 + H_2O$

38.
$$CH_3CH(NH_2)COOH \text{ (polymerized)} \rightarrow CH_3CH(NH_2)\overset{\overset{\displaystyle O}{/\!/}}{C} - NH - CH(CH_3)COOH, \text{ etc.}$$

39. $(C_{11}H_{23}COO)_3C_3H_5 + 3HOH \rightarrow C_3H_5(OH)_3 + 3C_{11}H_{23}COOH$

40. $2CH_3CH_2OH + 2K \rightarrow 2CH_3CH_2OK + H_2$

41. $CH_3CHOHCH_2CH_3 + CH_3CH_2COOH \rightarrow CH_3CH(OOCCH_2CH_3)CH_2CH_3 + H_2O$

42. $CH \equiv CCH_2CH_3 + HCN \rightarrow CH_2 = C(CN)CH_2CH_3$

43. $C_6H_6 + 3Cl_2$ (in 6 steps, with ultraviolet light as a catalyst) $\rightarrow C_6H_6Cl_6$ (hexachlorocyclohexane). This reaction proceeds with difficulty beyond the second step.

44.
$$CH_2(NH_2)COOH \text{ (polymerized)} \rightarrow CH_2(NH_2)\overset{\overset{\displaystyle O}{/\!/}}{C} - NH - CH_2COOH, \text{ etc.}$$

45. $CH_3CH(CH_3)CH_2CH_3 + Br_2 \rightarrow$ mixture of isomeric monobromopentanes

46.
$$CH_3CH(NH_2)COOH + CH_2(NH_2)COOH \rightarrow CH_3CH(NH_2)\overset{\overset{\displaystyle O}{/\!/}}{C} - NH - CH_2COOH \text{ (or reverse sequence)}$$

47. $CH_3CHOHCH_2CH_2CH_2CH_3 + HCl \rightarrow CH_3CHClCH_2CH_2CH_2CH_3 + H_2O$

48. $CH_3CH_2CH_2CHO + HCN \rightarrow CH_3CH_2CH_2\underset{\underset{\displaystyle CN}{|}}{CHOH}$

49. $CH_3CH_2CH_2COOH + CH_3CH_2CH_2CH_2OH \rightarrow CH_3CH_2CH_2COOCH_2CH_2CH_2CH_3 + H_2O$

50. $CH_3(CH_2)_4CHO + (O) \rightarrow CH_3(CH_2)_4COOH$

51. $CH_2 = C(CH_3)CH_2CH_2CH_3 + Br_2 \rightarrow CH_2BrCBr(CH_3)CH_2CH_2CH_3$

52. $C_6H_5OH + NaOH \rightarrow C_6H_5ONa + H_2O$

(continued)

53. $CH_2O + NH_3 \rightarrow H_2\underset{\underset{NH_2}{|}}{C}OH$

54. $C_3H_5(OH)_3 + 3CH_3CH_2CH_2COOH \rightarrow C_3H_5(OOCCH_2CH_2CH_3)_3 + 3H_2O$

55. $2C_6H_5OH + CH_2O \rightarrow (\ — C_6H_4OH — CH_2 — C_6H_4OH — CH_2 —)x$

56. $CH_3CH_2CH_2CH_2CHO + HCl \rightarrow CH_3CH_2CH_2CH_2CHCl(OH)$

57. $CH_3CH = CHCH_2CH_2CH_3 + HCl \rightarrow CH_3CH_2CHClCH_2CH_2CH_3$ and
    $CH_3CHClCH_2CH_2CH_2CH_3$

58. $C_6H_6 + H_2SO_4 \rightarrow C_6H_5SO_3H + H_2O$

59. $3(CH_3)_3COH + PCl_3 \rightarrow 3(CH_3)_3CCl + P(OH)_3$

60. $CH_3CH_2\overset{\overset{O}{||}}{C}\ CH_2CH_3 + H_2 \rightarrow CH_3CH_2CHOHCH_2CH_3$

61. $2CH_3OH + 2K \rightarrow 2CH_3OK + H_2$

62. $CH_3CH_2CHO + H_2 \rightarrow CH_3CH_2CH_2OH$

63. $CH_3(CH_2)_4CH_3 + Br_2 \rightarrow$ mixture of 1, 2, 3 monobromohexanes

64. $C_6H_6 + HNO_3 \rightarrow C_6H_5NO_2(+HNO_3)$ $C_6H_4(NO_2)_2$ (meta form)

65. $CH_3(CH_2)_6CH_3 + F_2 \rightarrow$ mixture of 1, 2, 3, 4 monofluorooctanes

Quantum Mechanics (pp. 141–142)

Group I.

1. 6.00×10^{14} Hz; 3.98×10^{-19} J
2. 5.0×10^{-7} m; 3.98×10^{-19} J
3. 6.8×10^{14} Hz; 4.4×10^{-7} m
4. 3.53×10^{14} Hz; 2.34×10^{-19} J
5. 6.7×10^{-6} m; 3.0×10^{-20} J
6. 1.30×10^{16} Hz; 2.31×10^{-8} m
7. 2.4×10^{14} Hz; 1.59×10^{-19} J
8. 3.65×10^{-8} m; 5.45×10^{-18} J
9. 1.84×10^{13} Hz; 1.63×10^{-5} m
10. 8.45×10^{14} Hz; 5.60×10^{-19} J
11. 8.923×10^{-7} m; 2.292×10^{-19} J
12. 5.436×10^{16} Hz; 5.519×10^{-9} m

Group II.

1. –1310 kJ/mole
2. For $n = 1$; $E_n = -313.6$ Kcal/mole
   For $n = 2$; $E_n = -78.40$ Kcal/mole
   For $n = 3$; $E_n = -34.84$ Kcal/mole
3. $E = 40.9$ Kcal/mole
4. $E = -58.8$ Kcal/mole
5. $f = 1.0 \times 10^{14}$ Hz
6. $E = 2.98 \times 10^{-12}$ ergs
7. 1.79×10^{12} ergs/mole
8. 4.14×10^{14} Hz
9. $r_3 = 9r_0$
10. $\lambda = \dfrac{h}{mc}$

Electrochemistry (pp. 143)

1. 54 g of silver
2. 82 g of zinc
3. 0.42 g of aluminum
4. 2.48 A

5. (a) +1.02 V; Fe^{3+} (aq) I Fe^{2+} (aq)
   (b) +0.46 V; Ag^+ (aq) I Ag (s)
   (c) +1.56 V; Ag^+ (aq) I Ag (s)
   (d) +0.74 V; Cu^{2+} (aq) I Cu (s)
   (e) +1.22 V; Br_2^0 (1)I Br^- (aq)
   (f) +2.00 V; Cu^{2+} (aq) I Cu (s)
   (g) + 0.49 V; Ni^{2+} (aq) I Ni (s)
   (h) +0.12 V; Co^{2+} (aq) I Co (s)

Index

The following list identifies the locations of exercises and tables for topics included in this book.